SYSTEM SAFETY FOR
THE 21ST CENTURY

SYSTEM SAFETY FOR THE 21ST CENTURY

THE UPDATED AND REVISED EDITION OF SYSTEM SAFETY 2000

Richard A. Stephans, PE, CSP
ARES Corporation

WILEY-INTERSCIENCE

A JOHN WILEY & SONS, INC., PUBLICATION

Library of Congress Cataloging-in-Publication Data is available.

ISBN 0-471-44454-5

Printed in the United States of America.

10 9 8 7 6 5

CONTENTS

FOREWORD TO *SYSTEM SAFETY FOR THE 21ST CENTURY*

I just heard it again. A colleague of mine said that he has always taken the "systems view" with regard to system safety. I was once again surprised, shocked is probably a better word, that not everyone had that view. It reminded me that there remain varying views of the scope of system safety. The scope of the system safety discipline is broad, just like the industries that use the discipline. The system safety discipline has expanded well beyond the U.S. Department of Defense community and U.S. borders and, as such, its recognized discipline approach and broad scope are becoming better define.

The System Safety Society and most system safety professionals take a broad view of the scope of system safety, a "system view." It considers the system safety discipline as analyzing all safety aspects for any size system (with a product being just a small system) throughout its entire life cycle. It uses a disciplined systems approach to manage safety risk by tapping into the known knowledge bases and using specific tools and techniques for analysis where knowledge bases do not exist or are insufficient for the technologies used in the system. Known knowledge bases include existing safety codes, safety standards, and lessons learned that have been developed in all technology areas. The system safety professional focuses more attention, however, where there are nonexistent or insufficient knowledge bases from which to draw upon. In this case, the system safety professional uses the specific tools and techniques available in the system safety profession to augment the lack of information in existing knowledge bases. The top-level analyses identify where new safety requirements are necessary and where existing safety codes and standards can be used. The system safety discipline bridges the gap when existing knowledge bases are lacking and manages safety risks by identifying hazards from the known knowledge bases and the tools and techniques of this profession.

Because the system safety professional focuses more attention where there are no or insufficient knowledge bases, some in industry perceive that the scope of the system safety discipline is just in those areas, where little or no knowledge bases exist. However, the scope of the system safety discipline is much broader and the system safety professional must have a complete understanding of how to use and apply the existing safety resources, in addition to when to use other system safety analyses to evaluate the entire system throughout its entire life cycle. Some colleagues refer to system safety as the "umbrella" safety, since you must draw upon all safety resources for the technologies involved in the design. The system safety discipline has an established methodology and unique tools for analysis. It establishes acceptable levels of risk as part of the process and does not necessarily seek zero risk or rely only on checklists or standards. It considers rare events and life-cycle operations and analyzes both normal and abnormal circumstances. The discipline manages for success using training, independent assessments, management commitment, and lessons learned and it plans for failure by establishing emergency response procedures, graceful degradation, surveillance, and maintenance.

This system safety discipline is unique because it addresses the safety of an entire system and its operations using existing knowledge bases and, where knowledge bases are insufficient, the tools of this profession. I am of the opinion that the methodology and tools of the system safety discipline should be applied to every system. I believe every company should develop and implement a system safety program that addresses the hazards in its organization, the products it purchases, and the systems that it designs and operates. Only the degree and depth of the system safety program will vary from system to system. As one colleague stated, I wouldn't spend too much time on the analysis of a paper clip. Using the system safety discipline, I am convinced that a company will apply its resources more effectively and achieve success in its ability to effectively manage safety risks.

The second edition of this book not only updates the text with the current information on standards such as MIL-STD-882D, it also adds another important tool and approach for the system safety engineer: a discussion on process safety in the chemical industry. Dick Stephans provides in-depth information of how to apply the system safety process to this specialized discipline: the users, distributors, or manufacturers of hazardous chemicals and related materials such as flammables and explosives. Historical accidents have demonstrated the need for legislation and specific legislative requirements from the Occupational Safety and Health Administration (OSHA) and the Environmental Protection Agency (EPA) are presented along with examples to reinforce understanding. Dick Stephans highlights the value of the system safety philosophy, in this case, to the chemical process standards and the application of methodologies to satisfy those requirements.

It is common now to see the application of the system safety approach, tools, and techniques in more and more industries without using the words *system safety*. This is evident by the more than 100 techniques described in the *System Safety Analysis Handbook*. While I am thrilled that the philosophy continues to expand, it is important to understand the basis for which most of the techniques are derived to ensure that they are applied appropriately.

Past President, System Safety Society (1999–2001)   PAIGE V. RIPANI

Professional credentials or experience in "systems safety" are not required to appreciate the potential value of the systems approach and system safety techniques to general safety and health practice. This book will help the reader move from system safety practice into far broader applications.

A joint conference of safety practitioners, led by the System Safety Society chapter in Washington, D.C., did much to expose the full capabilities of the systems approach to safety. The meeting produced a list of more than thirty techniques and approaches for use in system safety that were fully covered in the *Journal of the System Safety Society*. At least three interesting points emerged:

1. Only a few of the techniques were in regular use by system safety specialists.
2. Most of the techniques were in regular or partial use by members of the safety and health community who did not consider themselves system safety specialists or practitioners.
3. Most techniques had proponents who were not particularly receptive to other techniques. These backers were thus stakeholders in, and defenders of, a particular approach.

Bringing new ideas into the system is not easy, even if the ideas are good and people believe in them. They can be forced into practice, as the government has done on defense and certain other contracts. However, believers in the complete systems approach must also be able to convert their organizations to the idea. Few safety and health practitioners have the clout or skill to arrange this conversion.

A few system safety disciples and at least one government agency and one private group saw that no single approach leads to the level of safety performance needed for their complex operations. However, their ideas are not widely seen as having solid application in routine industrial safety and health practice. As a holistic approach emerged as a solution to long-range safety and health success, a few authors tried to place this complete approach into writing for the average practitioner. Their success was not spectacular, even when the material made good reading. The job of joining a holistic approach is harder because of the vested interests of various stakeholders and their approaches to safety and health problems. This book does not cast doubt on any of the viewpoints, but it does explore seldom-covered relationships that help us resolve their use for ourselves.

We find that the systems approach, old as it is, now figures prominently in most safety and health approaches and techniques. However, few system safety practitioners consider themselves as working in health fields such as stress management, wellness, industrial hygiene, or toxicology. Nevertheless, the fields are closely related to total practice. I have just reviewed the writings of two prominent industrial hygienists and a health physicist. Their success stems from viewing the whole system and any interacting systems—an interdisciplinary approach. Each of the three heads a major corporate safety and health department with "system safety" specialists. These three do not consider themselves system safety specialists but are wonders at applying a systems approach to their work.

One difficulty in applying certain systems approaches and techniques to problem solving is an inability of the practitioners to merge the various approaches and techniques, to relate them to each other, and to understand the relationship of diverse system safety techniques. Joe Stephenson shows in this text not only how the approaches vary, but also how they are similar and can interact with each other. This is a valuable service to the many disciplines and practitioners of the safety and health community.

Ranging from the traditional views of early systems safety adherents and developers, through the complete viewpoint of large-scale practitioners such as Idaho's System Safety Development Center to the all-encompassing viewpoint of DeBono, Stephenson brings it all into perspective. He relates how those tasks are visualized and traditionally used by system safety practitioners. He demonstrates how some of the systems approaches interface with each other and what they mean to their mutual success. Finally, he has made clear how some systemic techniques interface and can combine to form a complete system to solve safety and health problems.

Joe Stephenson makes practical the application of system safety techniques to safety and health problems not previously amenable to system safety solutions. Seeing the forest instead of the trees is a unique contribution of this book. The interaction of many disciplines and specialties can be seen. This book is a common ground for assessing a systems approach to safety and health disciplines and practice.

TED FERRY

As we continue into the twenty-first century, many challenges face the safety, engineering, and management communities. Risks and the potential for catastrophic loss are dramatically increasing as technology advances at an ever-increasing rate. The public demands a high level of safety in products and services, yet, in the face of world competition, the safety effort must be timely and cost-effective.

System safety tools and techniques currently used primarily in the aerospace, weapons, and nuclear industries offer great potential for meeting these challenges. The systematic application of system safety fundamentals early in the life cycle to produce "first time safe" products and services can provide significant, cost-effective gains in the safety effort in transportation, manufacturing, construction, utilities, facilities, and many other areas.

Yet, there are obstacles hampering current system safety efforts and restricting the expansion of system safety.

System safety continues, in many cases, to be more of an art than a science. The quality of system safety products is determined by the skill and talent of the individual analyst, not by the systematic application of accepted tools and techniques.

There is also a shortage of system safety engineers and of safety professionals, engineers, and managers trained in system safety.

A key factor is the lack of commonality of system safety terms, tools, and techniques.

The purpose of this book is to aid in expanding and improving the system safety effort to meet the needs of the next century by providing a basis for planning, evaluating, upgrading, conducting, and managing system safety programs.

It is designed to be used as a textbook, a planning guide, and a reference. This book is specifically written for:

- Safety professionals, including people in industrial and occupational safety, system safety, environmental safety, industrial hygiene, health, occupational medicine, fire protection, reliability, maintainability, and quality assurance
- Engineers, especially design engineers and architects
- Managers and planners
- Students and faculty in safety, engineering, and management

Students and others generally unfamiliar with system safety should read it straight through, in order, and retain it as a reference.

Managers and planners may find skimming through Part 1 first helpful, but will benefit most from Part 2.

Experienced system safety professionals are encouraged to keep an open mind—some will initially view parts of the book as heresy!—and be patient. A large portion of the book will be old hat to many of you, but several new concepts, techniques, and approaches are presented. Current practitioners may benefit most from Part 3.

Part 4 and the appendices contain how-to and reference information that should be of value to all who are interested in the system safety effort.

Part 5 is a new part devoted to process safety and particularly the U.S. OSHA and EPA rules to provide for safety to workers, the public, and the environment for those sites using certain hazardous substances above a listed threshold quantity. Most important is that that the level of calculated risk provides sites with a roadmap for safety actions.

Part 6 provides a discussion of professionalism that is important reading for the student and practitioner as well. The focus is on the system safety professional, but much of the information pertains to other related environmental, health, and safety fields.

A concerted effort was made to present information in a useful, clear, systematic, and understandable manner, with an emphasis on practical applications.

In summary, managers, engineers, and safety professions—regardless of previous system safety knowledge—should benefit from this book, with students and others unfamiliar with system safety learning the most and those applying the knowledge benefiting the most.

There are several people who either directly or indirectly helped or inspired the update to *System Safety 2000*. The following are just some of those people:

Joe Stephensen—THE author and teacher of system safety. We will miss him and his contribution to system safety.

Paige Ripani, past national president of the System Safety Society, acknowledged not just for her foreword to the 2nd edition, but also for more than 15 years of dedication to the field of system safety.

Pat Clemens is the unsung hero of system safety and risk analysis. Inspiration to many current (and future) system safety practitioners; former president of the Board of Certified Safety Professionals (BCSP).

Roger Brauer, the potentate of safety professionalism. We are indeed fortunate to have him in our midst. He has personally led a crusade to enhance the safety profession and the standard for safety professionals.

Paul Kryska—Leader and Manager of System Safety; National President of the Society at the time of publishing. Paul has vehemently practiced system safety in the Washington, D.C., area, in Albuquerque, NM, and now in Silicon Valley.

Warner Talso is the conscience of the System Safety Society and has been for more than ten years. He is the editorial power behind the publication of the first two editions of the *System Safety Analysis Handbook*. He is a best friend, a confident, and former Army nuclear weapons officer. I'll miss our Saturday breakfast burritos since my wife and I have moved to Nevada.

Perry D'Antonio of Sandia National Laboratories—the person who turned the Society around in 1995 and 1996.

Curt Lewis—International Society of Air Safety Investigators' Fellow and fellow director, BCSP. His daily Air Safety Bulletin is provided to thousands.

Fred Manuele, who provided the advice to "keep it a primer," whose guidance during the development of the current edition of the book provided a theme upon which this edition was structured.

Major Bob Baker, "Mr. Air Force System Safety," at the U.S. Air Force Safety Center at Kirtland AFB in New Mexico.

Michael Wilson and Pat McClure of the Los Alamos National Laboratory's D-5, Nuclear Design and Risk Analysis Group, who are "leading the world in risk analysis" and also providing key support in and beyond the United States for security and nuclear power safety.

To my employer, ARES Corporation, a relatively small, highly specialized, and highly respected company where everyone learns and provides excellence to its clients. They have been a repeat sponsor of the International System Safety Conferences and a technical power in government and industry risk assessment.

Finally, to my wife and most fervent supporter, Jo, who allowed me to add this "volunteer" project to my plate in the midst of family, work, Board of Certified Safety Professionals activities, and System Safety Society obligations.

ACKNOWLEDGMENTS FOR *SYSTEM SAFETY 2000*

I would like to thank three groups, all of whom contributed to *System Safety 2000*, albeit in different ways.

First, I would like to thank those who made direct contributions to the effort:

1. Ted Ferry, for graciously tolerating harassment during his well-earned retirement first to review the proposal for the book and later to write the foreword.
2. Bill Johnson, also in retirement, for his review of the proposal and for initial development of the MORT approach to system safety.
3. Randy Nason and the C. H. Guernsey Company of Oklahoma City (C. N. Stover, Jr., president) for the opportunity to prepare the FMEA and FTA examples found in Chapters 14 and 15, respectively, and for permission to use them and the generic preliminary hazard analyses included as Appendix D.
4. Bob Murray and Webb, Murray and Associates, Inc. (WMA) of Houston for permission to use materials developed while I was working for WMA.
5. Patsy Day of WMA for her assistance in preparing most of the graphics and course materials taught for WMA. These materials provided a significant input to *System Safety 2000*.
6. Kelly Seidel, for use of his personal library, resource materials, and expertise while I was researching, organizing, and writing the manuscript. His input, advice, and moral support throughout the project were invaluable, as was his assistance in performing our "real jobs."
7. All of the individuals who took the time and effort to respond to my questionnaires and to provide information found in the appendices.

Next, I would like to thank the individuals and organizations for and with whom I have worked during the last decade who have shared knowledge and afforded me the opportunity to learn, teach, and apply a variety of system safety tools on a variety of projects.

They are, in chronological order:

1. Reynolds Electrical and Engineering Company (REECo), an EG&G Company, Las Vegas. Special thanks to Collin Dunnam, Manager, Occu-

pational Safety and Fire Protection, and the exceptional staff of safety professionals. While responsible for system safety for REECo at the Nevada Test Site, I was given the opportunity to apply system safety tools and techniques to projects in support of the nuclear weapons testing program.

2. System Safety Development Center (SSDC), EG&G Idaho, Idaho Falls, Idaho, Bob Nertney, director (at that time), and the instructional staff, particularly Dick Buys (now with Los Alamos National Laboratory). While serving as a satellite instructor for the System Safety Development Center, I had the opportunity to teach MORT-based system safety and to interact with the SSDC staff and the Department of Energy and DOE contractor safety community.

3. National Safety Council, Chicago, Carl Piepho, Manager, Safety Training Institute. Carl provided me with the opportunity to teach MORT-based courses worldwide to the USAF ground safety community and to teach professional development seminars (most on system safety) annually at the National Safety Congress.

4. Webb, Murray and Associates, Inc. (WMA), Houston, particularly Bob Webb, Bob Murray, and Billy Magee, officers, and the talented WMA safety engineers and consultants. My time as director of WMA's Center for Advanced Safety Studies provided me with an opportunity to develop and teach system safety courses for NASA, DOD, DOT, and private industry and to participate in system safety projects.

5. From the U.S. Army, Don Pittenger, U.S. Army Corps of Engineers, and Paul Dierberger, U.S. Army Safety Center (also Harris Yeager, USAF; Craig Schilder, Naval Facilities Command; and Judy Sicka, U.S. Coast Guard) for the opportunity to develop and teach (through Kingsley Hendrick and the Department of Transportation's Transportation Institute and WMA) the facility system safety course.

Finally, I would like to thank my family for the tolerance, support, and understanding provided during the weekends, holidays, and early morning hours when I was hibernating in my office agonizing over a missed deadline. Special thanks to my wife, Phyllis, for her typing, copying, and mailing services and for extraordinary patience. And a sincere apology to my family for all the things we did not do in 1989 and 1990.

PART I

INTRODUCTION TO SYSTEM SAFETY

CHAPTER 1

The History of System Safety

Prior to the 1940s, safety was generally achieved by attempting to control obvious hazards in the initial design and then correcting other problems as they appeared after a product was in use or at least in a testing phase. In other words, designers relied, at least in part, on a trial-and-error methodology. In the aviation field, this process became known as the fly-fix-fly approach. An aircraft would be designed using the best knowledge available, flown until problems were detected (or it crashed), and then the problems would be corrected and the aircraft would be flown again. This method obviously worked best with low, slow aircraft.

That this approach was not acceptable for certain programs—such as nuclear weapons and space travel—soon became apparent, at least to some. The consequences of accidents were too great. Trial-and-error and fly-fix-fly approaches were not adequate for systems that had to be first-time safe.

Thus, system safety was born, or, more accurately, evolved. The history of system safety consists of

- Traditional trial-and-error or fly-fix-fly approach not adequate for aerospace and nuclear programs
- 1960s—MIL-STD-882 (DOD, NASA)
- 1970s—MORT (Department of Energy)
- 1980s—Other agencies

The roots of the system safety effort extend back at least to the 1940s and 1950s. Accurately tracing the early transition from the traditional trial-and-error approach to safety to the first-time safe effort that lies at the heart of system safety is really impossible, but such a transition occurred as both aircraft and weapon systems became more complex and the consequences of accidents became less acceptable.

System Safety for the 21ˢᵗ Century: The Updated and Revised Edition of System Safety 2000,
by Richard A. Stephans
ISBN 0-471-44454-5   Copyright © 2004 John Wiley & Sons, Inc.

THE 1960s—MIL-STD-882, DOD, AND NASA

Even though the need for a more in-depth, upstream safety effort was recognized relatively early in the aviation and nuclear weapons fields, not until the 1960s did system safety begin to evolve as a separate discipline. In the 1960s

- USAF publishes "System Safety Engineering for the Development of Air Force Ballistic Missiles" (1962)
- USAF publishes MIL-S-38130, "General Requirements for Safety Engineering of Systems and Associated Subsystems and Equipment" (1963)
- System Safety Society founded (1963)
- DOD adopts MIL-S-38130 as MIL-S-38l308A (1966)
- MIL-S-381308A revised and designated MIL-STD-882B, "System Safety Program Requirements" (1969)

Most agree that one of the first major formal system safety efforts involved the Minuteman intercontinental ballistic missile (ICBM) program. A series of pre-Minuteman design-related silo accidents probably provided at least part of the incentive (U.S. Air Force 1987).

Early system safety requirements were generated by the U.S. Air Force Ballistic System Division. Early air force documents provided the basis for MIL-STD-882 (July 1969), "System Safety Program for Systems and Associated Subsystems and Equipment: Requirements for." This document (and revisions MIL-STD-882A and MIL-STD-882B) became, and remain, the bible for the Department of Defense (DOD) system safety effort (Moriarty and Roland 1983).

In addition to weapon systems, other early significant system safety efforts were associated with the aerospace industry, including civil and military aviation and the space program.

Even though the National Aeronautical and Space Administration (NASA) developed its own system safety program and requirements, the development closely paralleled the MIL-STD-882 approach and the DOD effort, primarily because the two agencies tend to share contractors, personnel, and, to a lesser degree, missions.

Also, through the early to mid-1960s, the System Safety Society emerged. This professional organization was founded in the Los Angeles area by Roger Lockwood. Organizational meetings were held in 1962 and 1963. The organization was chartered as the Aerospace System Safety Society in California in 1964. The name was changed to System Safety Society in 1967 (Medford 1973). In 1973, the System Safety Society was incorporated as "an international, nonprofit, organization dedicated to the safety of systems, products, and services" (System Safety Society 1989).

THE 1970s—THE MANAGEMENT OVERSIGHT AND RISK TREE

In the late 1960s, the Atomic Energy Commission (AEC), aware of system safety efforts in the DOD and NASA communities, made the decision to hire William G. Johnson, retired manager of the National Safety Council, to develop a system safety program for the AEC.

In the mid-1970s AEC was reorganized into the Department of Energy (DOE). Even though the individual AEC programs and the AEC contractors had good (some better than others) safety programs in place, the programs and approaches varied widely. This lack of standardization or commonality made effective monitoring, evaluation, and control of safety efforts through-out the organization difficult, if not impossible.

Thus the goals of the AEC effort were to improve the overall safety effort by:

Developing a new approach to system safety that incorporated the best features of existing system safety efforts

Providing a common approach to system safety and safety management to be used throughout the AEC and by AEC contractors

In 1973 a revised management oversight and risk tree (MORT) manual was published by the AEC. Even though Johnson borrowed heavily from existing DOD and NASA programs, his MORT program bore little resemblance to programs based on MIL-STD-882 (Johnson 1973).

The work by Bill Johnson was expanded and supplemented throughout the 1970s by the System Safety Development Center (SSDC) in Idaho Falls, Idaho. The MORT program provides the direction for this second major branch of the system safety effort.

Progress in the 1970s included

- NASA publishes NHB 1700.1 (V3), "System Safety" (1970)
- AEC publishes "MORT—The Management Oversight and Risk Tree" (1973)
- System Safety Development Center founded (1974)
- MORT training initiated for AEC, ERDA, and DOE (1975)
- MIL-STD-882A replaces MIL-STD-882 (1977)

THE 1980s—FACILITY SYSTEM SAFETY

Throughout the 1980s, three factors have driven system safety tools and techniques in areas other than the traditional aerospace, weapons, and nuclear fields.

First, the complexity and high cost of many nonflight, nonnuclear projects have dictated a more sophisticated upstream safety approach. Second, product liability litigation has provided added incentive to produce safe products, and, third, system safety experience has begun to demonstrate that upstream safety efforts lead to better design. System safety tools and techniques originally considered to be expensive but necessary add-ons have proven to be cost-effective planning and review tools.

Significant programs initiated or developed in the 1980s include the facility system safety efforts of the Naval Facilities Command and the U.S. Army Corps of Engineers and initiatives in the petrochemical industry.

- MIL-STD-882B replaces MIL-STD-882A (1984)
- NAVFAC sponsors system safety courses (1984)
- AIChE publishes "Guidelines for Hazard Evaluations Procedures" (HazOps) (1985)
- MIL-STD-882B updated by Notice 1 (1987)
- USACE-sponsored facility system safety workshops initiated (1988)

The need for a system safety effort for major military construction projects resulted in the development of draft guidelines and facility system safety workshops for the military safety and engineering communities. By the end of the decade, facility system safety training programs for government employees were established, and similar courses for contractors were available. Regulations outlining facility system safety efforts were pending, and facility system safety efforts were being required on selected military construction projects. In addition, NASA was initiating facility system safety efforts, especially for new space station support facilities.

In 1985, the American Institute of Chemical Engineers (AIChE) initiated a project to produce the "Guidelines for Hazard Evaluation Procedures." This document, prepared by Battelle, includes many system safety analysis tools. Even though frequently identified as hazard and operability (HazOp) programs, the methods being developed by the petrochemical industry to use preliminary hazard analyses, fault trees, failure modes, effects, and criticality analyses, as well as similar techniques to identify, analyze, and control risks systematically, look very much like system safety efforts tailored for the petrochemical industry (Goldwaite 1985).

THE 1990s—RISK-BASED PROCESS SYSTEM SAFETY

If the 1980s was designated as "facility safety," then the 1990s should be identified as "process safety." Prior to the 1990s, OSHA regulations were almost exclusively compliance-based. Very specific rules were promulgated and inspections were made to ensure that the rules were followed. The OSHA process safety regulation (29 CFR 1910.119) required that the risk associated with a manufacturing or chemical processing site with listed substances be assessed and appropriate actions be taken to mitigate the results of an accident to protect the workers.

In addition, there was greater interface with quality assurance (QA) to include the management of change segment so important to safety. An analogy related to the safety aspect of change management is advice given when first driving a car. As long as you are in the same lane and stay there, you are fairly safe, but when you have to move lanes or make a turn, you must be particularly careful and watchful about what you are about to do. Much the same can be said about changes to hardware and software during design, development, and fielding. It also became more apparent that the "quality" of input materials was very important to desired output product as well as to the safety performance of the product. Product impurities could weaken a structure and result in an undesired chemical reaction with intermediate chemical ingredients.

Further, the QA audit function is directly related to safety. QA audits with an encompassing scope include facility and process safety as one of the elements reviewed.

Milestone Standards Issue Events

System safety related events and guidance documentation evolution and emergence in the 1990s included: 882C in 93; the *System Safety Analysis Handbook* in 1993 (with a second edition in 1997). The Handbook is currently sold in more than 35 countries; PSM in 1992; RMP in 1996; *Hazard Prevention* changed to the *Journal of System Safety* in 1999; European Machining Standard, European Norm (EN) 1050 in 1997 that requires risk analysis prior to mechanical or electrical controls; the System Safety Society increases frequency of international conferences to annually; Center for Chemical Process Safety established; publication of the "System Safety and Risk Management—NIOSH Instructional Manual" by Dr. Rodney Simmons and Pat Clemens, both strong advocates of system safety and both very closely tied to the maturation of the Board of Certified Safety Professionals.

At the 1993 International System Safety Conference, the then Chief of Air Force Safety in his keynote presentation said that the two challenges for the 90s were in software system safety and in human factors. This was true then, and it is today, more than ten years later.

THE 2000s—QUEST FOR INTRINSIC SAFETY

As we progress into the new century there are both opportunities and challenges. Opportunities present themselves in the form of (1) the potential of integrating system software safety with control engineering to more closely achieve a level of "intrinsic safety" and (2) the proliferation of system safety as a discipline in other parts of the world. The challenges we face include the realization of security needs after the terrorist attack on the United States on September 11, 2000. Additional challenges and a future prediction are presented in Chapter 5.

We define intrinsic safety as safety designed and built into a system. Yes, this is an overlap with system safety. The two concepts are converging. Intrinsic safety is certainly a noble goal and one that should be continually pursued.

The proliferation of information on the Internet can be overwhelming. It is important to know how to move around the Internet using search engines, generic sites, and links. The Internet can provide much useful information in a relatively short period of time.

Milestone Standards Issue Events

MIL-STD-882D (10 FEB 2000); ANSI B11.TR3 (2000) which promulgated guidance related to safety through design; EN1050 (97) for risk assessment prior to the installation of new machinery; the U.S. Department of Energy's 10 CFR 830 in 2001, effective in 2003 for nuclear safety management.

A new issue of Military Standard 882 in the year 2000 provided for a modified approach to system safety achievement. The Standard was renamed, as a "Standard Practice for System Safety" and its implementation is further discussed in Chapter 3 "Current Approaches to System Safety." Suffice to say here, the standard allows for flexibility in implementation while preserving basic system safety requirements.

REVIEW QUESTIONS

1. Explain the fly-fix-fly approach to developing safe products.
2. What made the fly-fix-fly approach unacceptable? What were the first types of programs to seek something better than the fly-fix-fly approach?
3. How did MIL-STD-882B evolve? Where did it originate and who uses it now?
4. How did the MORT approach to system safety develop? Who developed it and who uses it?
5. Identify the three primary factors that drove the expansion of system safety during the 1980s.
6. Name three areas in which system safety efforts were initiated or developed during the 1980s.
7. Discuss why, although it is appropriate to transition to risk-based safety, it is not possible or appropriate to entirely do away with safety compliance.
8. Name two quality assurance program elements and explain their interface with system safety.

9. Name and explain why two locations outside the United States are becoming "centers" for safety standards development.

10. Although the Internet existed for a decade prior to the year 2000, why is it so much more popular today?

REFERENCES

Johnson, William G. 1973. *MORT, The Management Oversight and Risk Tree.* Washington, D.C.: U.S. Atomic Energy Commission.

Medford, Fred. 1973. History of the system safety society (SSS). *Hazard Prevention* 9(5):38–40.

Moriarty, Brian, and Roland, Harold E. 1983. *System Safety Engineering and Management.* New York: John Wiley & Sons.

U.S. Air Force. 1987. *SDP 127–1: System Safety Handbook for the Acquisition Manager.* Los Angeles: HQ Space Division/SE.

CHAPTER 2

Fundamentals of System Safety

BASIC DEFINITIONS

One of the major problems confronting the system safety community is a lack
of standardization or commonality. (This problem is discussed in detail in
Chapter 4.) Presenting "universally accepted" definitions to even basic terms
is therefore difficult because, by and large, they do not exist. The following
terms are defined in nontechnical language to ensure the reader understands
each term as used in this book. Specific definitions from documents widely
used in the system safety effort are contained in the glossary, and definitions
used by specific organizations are included in Chapter 3.

Safety: Freedom from harm. Safety is achieved by doing things right the first
time, every time.

System: A composite of people, procedures, and plant and hardware
working within a given environment to perform a given task (Fig. 2-1).

System safety: The discipline that uses systematic engineering and management techniques to aid in making systems safe throughout their life
cycles.

Hazard: Something that can cause significant harm.

Risk: The chance of harm, in terms of severity and probability.

Safety community: That group of individuals who provide staff support to
the line organization in support of the safety effort. It includes occupational and industrial safety, system safety, industrial hygiene, health,
occupational medicine, environmental safety, fire protection, reliability,
maintainability, and quality assurance personnel.

FUNDAMENTAL SAFETY CONCEPTS

Five fundamental safety concepts apply to any safety effort.

System Safety for the 21ˢᵗ Century: The Updated and Revised Edition of System Safety 2000,
by Richard A. Stephans
ISBN 0-471-44454-5   Copyright © 2004 John Wiley & Sons, Inc.

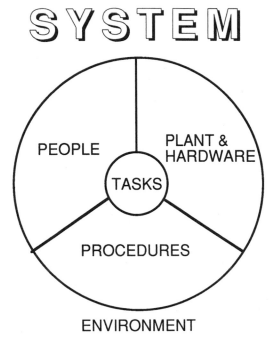

SYSTEM

ENVIRONMENT

Figure 2-1   System definition: people, procedures, and plant hardware performing specific tasks in a given environment.

1. Safety is a line responsibility.
2. Safety is productive.
3. Safety requires upstream effort.
4. Safety depends on the safety precedence sequence.
5. Systematic tools and techniques help.

Safety Is a Line Responsibility

Line managers and supervisors are responsible for the safety of their organizational units and operations. This old principle is extremely important in the safety world. This particular fundamental must be understood and accepted by all parts of the organization. The safety professional's job is to provide the staff support necessary to ensure that the line organization is able to do its job well and effectively. The system safety effort does not relieve program or project managers or design engineers of their safety responsibilities.

Safety Is Productive

Safety is achieved by doing things right the first time, every time. If things are done right the first time, every time, we not only have a safe operation but also an extremely efficient, productive, cost-effective operation. Because the system safety effort was founded to meet the requirement for first-time safe operations, the system safety effort obviously supports this principle.

Safety Requires Upstream Effort

The safety of an operation is determined long before the people, procedures, and plant and hardware come together at the work site to perform a given task.

> Effectively evaluating or significantly influencing the long-term success of an operation at the work site alone is virtually impossible.

The selection of personnel, the ongoing and initial training, the development of the procedures, and the design of the facilities and equipment are the types of tasks that ultimately determine the safety of the workplace (Fig. 2-2). Safety professionals, managers, or supervisors who think they can have a significant impact on safety in the workplace by putting on a hard hat and safety shoes and meandering out with a clipboard to make the world safe are really fooling themselves if the upstream processes have not been done properly. Good safety practices must begin as far upstream as possible.

Improvements in safety can often be made for a minimal amount of money if they are made far enough upstream. Sometimes the same changes may be extremely costly to the point of being impractical or even impossible if the potential hazards or the shortcomings in the system are not recognized until

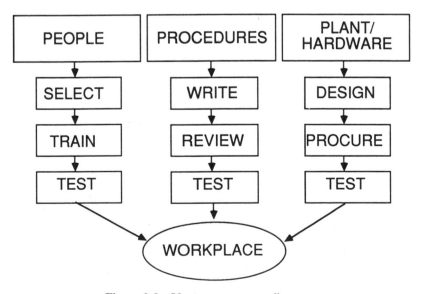

Figure 2-2   Upstream process diagram.

the system comes together in the workplace. Again, the system safety effort directly supports this safety fundamental.

Safety Depends on the Safety Precedence Sequence

The fourth fundamental involves something called the *safety precedence sequence*. It is a prioritized list of controls that should be considered and applied, in sequence, to eliminate or control identified hazards.

- Design for minimum hazard.
- Provide safety devices.
- Provide warning devices.
- Control with procedures and training.
- Accept remaining residual hazards.

The first and most effective way to control identified hazards is to eliminate them through design or engineering changes.

If controlling the hazards through improved design or engineering is impossible or impractical, the next course of action should be to use physical guards or barriers to separate potential unwanted energy flows or other hazards from potential targets.

Warning devices should next be applied to any remaining hazards.

As a last resort, after other methods have been exhausted, procedures and training should be used.

Even after all these controls have been applied, some residual risks may remain. Making most operations or systems totally risk-free is impossible. The residual risks should be identified and formally accepted by the appropriate level of management. If the residual risks are unacceptable, additional controls should be provided or the project should be abandoned.

For most operations, a combination of controls must be used. Correct application of the safety precedence sequence dictates that lower-level controls are not considered until all practical efforts have been exhausted at higher levels.

The safety precedence sequence originated as part of the system safety effort and is one of the few tools that is common to nearly all system safety efforts.

It is not, however, always called the *safety precedence sequence*. It is also known as the *system safety precedence*, the *hazard reduction precedence sequence*, the *hazard control precedence sequence*, the *risk control sequence*, and several other names. The fifth step (accept residual risks) is not always included.

Systematic Tools and Techniques Help

The use of systematic methods and a systematic approach to identifying, analyzing, and controlling hazards aids in reducing error rates and performing safety tasks more effectively and efficiently.

FIND THE "F'S"

Count the number of times the letter "F" appears in the following sentence.
Enter the number you find here:_____

FINISHED FILES ARE THE RESULT OF YEARS OF SCIENTIFIC STUDY COMBINED WITH THE EXPERIENCE OF MANY YEARS.

Figure 2-3   Find the "Fs" exercise.

To demonstrate this, take the simple test in Figure 2-3. Then take the test again, using the systematic method recommended in Figure 2-4. This test illustrates rather clearly that systematic methods can aid in reducing error rates for even simple, straightforward tasks.

SYSTEM SAFETY FUNDAMENTALS

What Is *System Safety*?

Simply put, *system safety* is the name given to the effort to make things as safe as is practical by systematically using engineering and management tools to identify, analyze, and control hazards.

The system safety "effort" is sometimes called an *approach*, a *discipline*, a *concept*, a *doctrine*, and/or a *philosophy*.

"Things" to be made safe can be systems, programs, projects, products, operations, or facilities.

"As safe as is practical" may be expressed as the "best degree of safety," "optimum safety," or "optimum risk management" within constraints (operational effectiveness, cost, and time).

Most formal definitions of *system safety* also emphasize that the effort is for "all phases of the life cycle," even though most efforts are concentrated (and rightfully so) in the early phases of new programs.

FIND THE "F'S"

Count the number of times the letter "F" appears in the following sentence by starting at the end and working backward, one letter at a time. Place an "X" on each "F", then count the X's.

Enter the number you find here:_____

FINISHED FILES ARE THE RESULT OF YEARS OF SCIENTIFIC STUDY COMBINED WITH THE EXPERIENCE OF MANY YEARS.

Figure 2-4   Find the "Fs" using a systematic method. Almost everyone finds only three or four *Fs* by simply reading through the sentence; almost everyone finds all six by using the simple, systematic method recommended.

Why Do We Do System Safety?

The primary reason for using a system safety approach is to achieve better safety. Most traditional safety programs are compliance oriented; that is, safety is achieved by complying with the appropriate codes, standards, and regulations. This approach has two shortcomings. First, codes, standards, and regulations are, by and large, political documents. As such, they are frequently the result of compromise and represent minimum acceptable levels of performance. Even though 100% compliance is a laudable and sometimes challenging goal, a program achieving 100% compliance meets only minimum standards. The system safety effort attempts to exceed minimum standards and to provide the highest level of safety achievable.

A second shortcoming of a safety program based only on compliance is that it tends, by nature, to be reactive. One of the oldest clichés in the safety world is that safety codes, standards, and regulations are "written in blood"; that is, most regulatory requirements were written in reaction to an accident or series of accidents. Thus, codes, standards, and regulations tend to lag behind rapidly changing technology so that a gap always exists between "compliance" safety and "optimum" safety. Codes, standards, and regulations tend to be inadequate for research and development and other leading-edge activities.

Compliance safety relies on

- Codes, standards, and regulations
- Engineer responsibility
- 100% compliance to minimum standards

Systems safety relies on

- Systematic hazards analyses
- Team effort
- Optimum safety

The system safety effort strives to be proactive by very early identification, analysis, and control of hazards to produce first-time safe systems.

How Do We Do System Safety?

Most major system safety programs to date have involved government acquisitions. Thus, "how" has tended to be a two-part question. First, how does the government develop system safety requirements, communicate these requirements to the appropriate contractors, and manage the effort to ensure that the requirements are met?

Second, how do contractors develop and implement a system safety program to perform the tasks necessary to meet the government requirements?

The government agency (or other procuring organization) for whom a system (or product) is being developed must determine the objectives and specifications for the project. The specifications should include standards of safety performance and define the levels of acceptable risk. The specific system safety tasks and requirements must also be defined.

These standards of safety performance and system safety requirements must then be communicated to potential contractors in prebid conferences, in the request for proposal (RFP), and/or in the statement of work (SOW) for the project. The RFP may require contractors to include in their proposals a plan for meeting the system safety requirements. These requirements are also written into the contract.

In addition to determining system safety tasks and standards, the government must also develop a plan to monitor and evaluate the program conducted by the contractor to ensure that requirements are being met. This objective is normally accomplished by reviewing progress and evaluating "deliverables." Safety reviews are normally linked to milestones in the overall project schedule. The deliverables are the physical documents that are to be generated and provided as system safety products.

The government plan for developing, communicating, and monitoring the system safety effort may be called the *system safety management plan.*

The plan developed by the contractor to meet the system safety program requirement of the government (as specified in the contract) is generally called the system safety program plan (SSPP).

The system safety program plan generally includes detailed information about system safety personnel (responsibilities, qualifications, and level of effort), procedures (tasks to be performed and techniques to be used), and products (format and schedule).

The primary objectives of the system safety effort are to identify, analyze, and control hazards to the extent possible with constraints of operational effectiveness, time, and money.

A typical system safety task for hazard identification would be the preparation of a preliminary hazard list (PHL). Hazard identification or discovery is accomplished by reviewing lessons learned, accident reports, and other historical data. A PHL may be prepared through an informal conference, the use of checklists, and occasionally other techniques such as energy trace and barrier analysis (ETBA).

Common hazard analysis tasks include the preparation of a preliminary hazard analysis, systems and subsystem hazard analyses, and an operating hazard analysis. These tasks also aid in the hazard control and hazard reduction effort.

The preliminary hazard analysis (PHA) is an initial look at the entire system. A PHL, if available, is expanded by adding new hazards that may be identified as more project information is developed, as well as more information about each hazard. If a PHL has not been prepared, the PHA serves as the primary hazard identification tool as well as the initial hazard analysis. The methods used for conducting a PHA are basically the same as for a PHL, even though occasionally more advanced techniques may be appropriate.

As a project is developed and more detailed design data are available, a system hazard analysis (SHA) and subsystem hazard analyses (SSHAs) may be conducted to provide more detailed, in-depth risk assessment information. Two of the more widely used techniques for performing SHAs and SSHAs are the failure modes and effects analysis (FMEA) and fault tree analysis (FTA).

Failure modes and effects analysis is a systematic look at hardware, piece by piece, to determine how each piece could fail. The effects of each type of failure on the surrounding pieces and on the system or subsystem as a whole, and an assessment of the risk associated with each failure, commonly in terms of severity and probability, are expressed as a risk assessment code (RAC).

In most organizations that have a "reliability" effort separate from the safety or system safety effort, an FMEA is considered a reliability tool. The "safety" version is called a failure modes and effects criticality analysis (FMECA).

Fault tree analysis is used to examine, in detail, how a specific unwanted event could occur. It is a deductive, top-down effort normally reserved for examining critical types of failures or mishaps. Fault tree analysis can be qualitative or quantitative.

The final major type of analysis is the operating hazard analysis (OHA) or the operating and support hazard analysis (O&SHA). The OHA is the analysis that integrates the people and procedures into the system. The PHA, and especially the SHA and SSHA, are almost exclusively focused on hardware. A relatively new tool, the project evaluation tree (PET), may be used for performing OHAs. A PET analysis uses an analytical tree systematically to evaluate the procedures, personnel, and hardware for a new or ongoing project.

In summary, the typical system safety products are

- System safety program plan (SSPP)
- Preliminary hazard analysis (PHA)
- Subsystem hazard analysis (SSHA)
- System hazard analysis (SHA)
- Operating hazard analysis (OHA)

In addition to the analytical techniques already mentioned, an ever-growing list of other techniques is available for performing hazard analysis (Chapter 17).

The common analytical techniques are

- Failure modes and effects analysis (FMEA)
- Fault tree analysis (FTA)
- Energy trace and barrier analysis (ETBA)
- Management oversight and risk tree (MORT)
- Project evaluation tree (PET)
- Change analysis
- Common cause analysis

A major objective of most of these analyses is to produce meaningful risk assessment data to aid in prioritizing hazards, allocating resources, and evaluating the acceptability of risks associated with these hazards. A common tool for assessing risk is the risk assessment code (RAC). The RAC is the number associated with a given level of severity and a given probability of occurrence as shown on a risk assessment matrix, with severity on one axis and probability on the other (Table 2-1).

Safety is achieved by continuing to use the safety precedence sequence to control hazards until all identified hazards are eliminated or controlled to an acceptable level and residual risks are formally accepted. The detailed procedures for hazard tracking and risk resolution should be included in the SSPP.

When Do We Do System Safety?

The system safety effort should begin when the project begins and continue throughout the life cycle. The system safety effort concentrates, however, on

TABLE 2-1. Risk Assessment Matrix

Frequency of Occurrence	Hazard Categories			
	I Catastrophic	II Critical	III Marginal	IV Negligible
FREQUENT	1	2	7	13
PROBABLE	2	5	9	16
OCCASIONAL	4	6	11	18
REMOTE	8	10	14	19
IMPROBABLE	12	15	17	20

Hazard risk index
 1–5: unacceptable
 6–9: undesirable
 10–17: acceptable with review
 18–20: acceptable without review

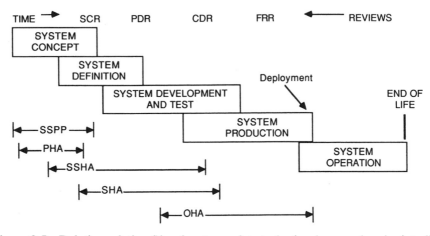

Figure 2-5   Relative relationship of system safety tasks (analyses and review) to life cycle phases for a simple one-of-a-kind system.

the upstream processes, with the PHL serving as a guide to initial program requirements. The SSPP outlines the system safety program, with the PHA being the first type of analysis, followed by the SHA or SSHA and the OHA (Fig. 2-5).

Generally, each analysis should be initiated as soon as appropriate design data are available and completed as soon as practical. Reviews should be spaced throughout the program, especially at program milestones and at the completion of each phase of development.

Who Does System Safety?

Basically, all major components of an organization have a role in system safety.

Management has the overall responsibility for ensuring that system safety programs are established, that they are adequately staffed at all levels, that the training is conducted for all personnel associated with the system safety effort, that safety concerns are identified and communicated, and that adequate resources are allocated.

The safety community has the responsibility for providing the staff support to know, teach, and apply the specific system safety tools and techniques and to help establish and monitor system safety programs.

The engineering community has the responsibility for providing hardware expertise to perform system safety analysis and to make necessary design changes.

At the working level, system safety tasks are normally performed by a system safety working group (SSWG). One of the reasons that a system safety working group or a team approach is generally used in system safety efforts is that multiple talents and disciplines are required in order to provide for the first-time safe, efficient operation of the entire system (Fig. 2-6).

Four specific groups, each bringing skills and knowledge vital to a good effort, are typically included in a SSWG.

Project management provides overall direction; management and organizational skills; project knowledge in the areas of constraints, goals, and objec-

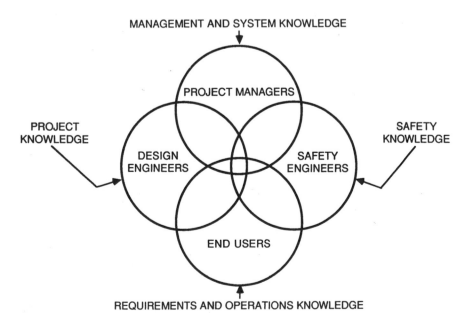

Figure 2-6 System Safety Working Groups.

tives; and decision-making authority. The project management representative chairs the SSWG and provides liaison with higher levels of management.

The safety community provides expertise in hazard identification, analysis, and control techniques. The safety representative may serve as the primary advisor to the chairperson in articulating system safety goals, tasks, and responsibilities. The safety representative frequently has the job of writing or drafting documents generated by the SSWG.

The engineering community provides specific project knowledge. The engineering representative(s) should ensure that safety concerns and hazards are addressed as early as practical in the design process.

The final group to be included is sometimes overlooked but is the key player in the design and development process: the end user. The representative from the organization that will use the end product provides detailed requirements and operations information.

The size of the system safety working group varies, depending on the scope and complexity of the project. System safety working groups may be required at several levels of the organization, with upper-level SSWGs providing primarily review and oversight. These upper-level teams may be referred to as *system safety groups* (SSGs) instead of SSWGs.

SYSTEM SAFETY TENETS

With the fundamentals of this chapter in mind and the experience of a career in the field, there are certain written and unwritten basics that are worth repeating because of their vital importance to the safety practitioner. These are the essential actions needed to perform the system safety undertakings:

1. Systematically identify, evaluate, and control hazards in order to prevent (or mitigate) accidents.
2. Apply a precedence of controls to hazards starting with their elimination, designing to preclude hazards, and finally administrative controls. Administrative controls include signs, warnings, procedures, and training. (The lowest precedence are those controls that rely on people.)
3. Perform proactively rather than reacting to events. This starts with a program plan.
4. Design and build safety into a system rather than modifying the system later in the acquisition process when any changes are increasingly more expensive.
5. Develop and provide safety-related design guidance and give it to the designers as the program is initiated.
6. Use appropriate evaluation/analysis techniques from the tabulated variety available.

7. Rely on factual information, engineering, and science to form the basis of conclusions and recommendations.

8. Quantify risk by multiplying the ranking of undesired consequences of an event by the probability of occurrence. There are variations to this "equation."

9. Design, when allowed, to minimize or eliminate single-point failures that have an undesired consequence. Make at least 2-fault tolerant, that is tolerant of multiple faults or system breakdown that would have adverse safety consequence.

10. Identify, evaluate, and control hazards throughout the system's life and during the various operational phases for normal and abnormal environments.

11. After application of controls to mitigate a hazard(s), management must recognize and accept the residual risk.

12. Recognize the quality assurance interface: (1) Decrease risk by using materials that are properly specified and possess adequate quality assurance and (2) implement to continually improve the system.

13. Tabulate and disseminate lessons learned and incorporate those lessons for future safety enhancement.

14. Apply system safety to systems to include processes, products, facilities, and services.

15. Recognize that near-miss conditions, if not corrected, most likely develop into accidents.

REVIEW QUESTIONS

1. List and discuss the four fundamentals that apply to any safety program.
2. Discuss the roles of the safety manager and the project manager in the overall safety effort.
3. State briefly how safety is achieved.
4. Explain the role of the upstream process in determining workplace safety.
5. Define and discuss the *safety precedence sequence*.
6. Define *system safety*.
7. List the three basic system safety tasks.
8. Discuss the difference between system safety and compliance safety.
9. List and define six typical system safety "products" or documents that a contractor may be required to prepare.
10. A system safety working group should have representatives from at least four groups or "communities." Name them and discuss the skills or knowledge that each representative brings to the system safety effort.

REFERENCES

U.S. Army. (undated). Facility System Safety Manual. Washington, D.C.: Hq, USACE. Draft.

U.S. Department of Defense. 1984 (updated by Notice 1, 1987). *MIL-STD-882B: System Safety Program Requirements*. Washington, D.C.: U.S. Department of Defense.

CHAPTER 3

Current Approaches to System Safety

DEPARTMENT OF DEFENSE

MIL-STD-882D

As this edition is published, the current MIL-STD-882 is the "D" version and, therefore, discussion of that approach to system safety implementation follows. The "D" version responds to changes in the defense acquisition procedures and is no longer the source for safety-related data item descriptions (the formal detailed requirements associated with defense procurement contracts). Therefore, for several reasons, the more detailed methods of implementation of previous versions of the military standard are retained because they are still valid. This is for two reasons. First, there are some government contracts with very long periods of performance or those that have been renewed multiple times that still refer to the earlier standard version. Secondly, and perhaps more importantly, the implementation method of the previous versions provides a guide for implementation of the newer "standard practice."

In a memorandum to the Defense Acquisition Community, the following was quoted about system safety

Department of Defense policy is to rely on performance based requirements whenever practicable and to not require standard management approaches or manufacturing processes in solicitations and contracts. By establishing performance requirements and then relying on contractors to meet those requirements, we dramatically reduce contractual call-out of specific specifications and standards and enable innovation. Most importantly, this allows contractors to meet our full needs—including safety considerations—at the lowest cost.

Discontinuing the practice of including documents in the contract which mandated standard approaches does not mean that we no longer care about the issues addressed by those documents. In fact, Defense program managers and contract oversight personnel must now have even greater understanding of the underly-

System Safety for the 21ˢᵗ Century: The Updated and Revised Edition of System Safety 2000, by Richard A. Stephans
ISBN 0-471-44454-5   Copyright © 2004 John Wiley & Sons, Inc.

ing management or engineering processes at work, and the results required, so that they can evaluate and monitor contractor processes designed to achieve the same ends under this more flexible approach.

Some individuals have construed our new acquisition philosophy to indicate that the Department has slacked off on system safety. Quite the contrary, DoD 5000.2-R, "Mandatory Procedures for Major Defense Acquisition Programs and Major Automated Information Systems," requires program managers to have an aggressive system safety program, and to continually work with their contractors to identify and mitigate design-induced safety risks. Many of the old specifications and standards that were used to constrain design in an attempt to ensure safety are being rewritten into handbooks providing guidance for designers, along with rationale and lessons learned. The end result is a more effective library of design information and guidance leading to safer designs at lower cost, and with greater flexibility given to the contractor to meet all of the performance requirements.

—(R. Noel Longuemare, U.S. Defense Department, Memorandum, *System Safety and Milspec Reform*, dated August 7, 1997.)

Where previous versions of the standard (particularly versions B and C) provided detailed requirements and specify how to implement the system safety program, the current version is focused more toward the way the acquisition contract is prepared. The current procedure is important because it provides some flexibility to the acquisition program manager to create a system safety program that fits the particular acquisition.

The MIL-STD-882D standard practice describes a system safety approach that is useful in the management of Environmental, Health of Safety mishap risks encountered in the life cycle of Department of Defense (DOD) systems, subsystems, equipment, and facilities. To paraphrase the standard, mishap risk must be identified, evaluated, and mitigated to a level acceptable (as defined by the system user or customer) to the appropriate authority, and compliant with federal laws and related rules. Further, residual mishap risk associated with an individual system must be reported to and accepted by the appropriate authority. These basic requirements are fundamental to system safety.

The standard has both general and detailed requirements. General requirements include documentation of the system safety approach, identification of hazards, risk assessment, identification of risk mitigation measures, reduction of risk to an acceptable level, verification of risk reduction, review of hazards and acceptance of residual risk, and tracking of hazards and residual risk. When a government contract specifies MIL-STD-882D and no other requirement, only the general requirements apply.

Specific requirements include a listing of 14 program manager responsibilities. These tend to be the basic rules for any project manager with a focus toward system safety.

The most common and widespread approach to system safety is that taken by the DOD and DOD contractors. They originally developed and implemented system safety programs. The DOD approach is based on MIL-STD-882B system safety program requirements.

Familiarity with MIL-STD-882B is necessary in order to understand DOD system safety programs. This document was officially issued on 30 March 1984 to supercede MIL-STD-882A (28 June 1977) and was updated by Notice 1 on 1 July 1987.

The purpose of MIL-STD-882B is stated as:

> This standard provides uniform requirements for developing and implementing a system safety program of sufficient comprehensiveness to identify the hazards of a system and to impose design requirements and management controls to prevent mishaps by eliminating hazards or reducing the associated risk to a level acceptable to the managing activity (MA). The term "managing activity" usually refers to the Government procuring activity, but may include prime or associate contractors or subcontractors who wish to impose system safety tasks on their suppliers.

The document provides important guidance and how-to information to be used in developing and imposing system safety requirements. It does not provide significant guidance on how to meet the requirements. Contractors have the responsibility for developing and implementing system safety program plans to meet the system safety program requirements imposed by the government and MIL-STD-882B.

The first few pages of the document contain such topics as the scope, referenced documents, definitions and abbreviations, and system safety requirements.

Key definitions include

Hazard: A condition that is prerequisite to a mishap.

Risk: An expression of the possibility of a mishap in terms of hazard severity and hazard probability.

Safety: Freedom from those conditions that can cause death, injury, occupational illness, or damage to or loss of equipment or property.

System: A composite, at any level of complexity, of personnel, procedures, materials, tools, equipment, facilities, and software. The elements of this composite entity are used together in the intended operational or support environment to perform a given task or achieve a specific pro-duction, support, or mission requirement.

System safety: The application of engineering and management principles, criteria, and techniques to optimize safety within the constraints of oper-ational effectiveness, time, and cost throughout all phases of the system life cycle.

The system safety requirements paragraph discusses program objectives, general design requirements, the system safety precedence (safety precedence sequence), and risk assessment descriptions of both severity and probability categories (Tables 3-1 and 3-2).

TABLE 3-1. Hazard Severity

Description	Category	Mishap Definition
CATASTROPHIC	I	Death or system loss
CRITICAL	II	Severe injury, severe occupational illness, or major system damage
MARGINAL	III	Minor injury, minor occupational illness, or minor system damage
NEGLIGIBLE	IV	Less than minor injury, occupational illness, or system damage

TABLE 3-2. Hazard Probability

Description*	Level	Specific Individual Item	Fleet or Inventory†
FREQUENT	A	Likely to occur frequently	Continuously experienced
PROBABLE	B	Will occur several times in life of an item	Will occur frequently
OCCASIONAL	C	Likely to occur sometime in life of an item	Will occur several times
REMOTE	D	Unlikely but possible to occur in life of an item	Unlikely but can reasonably be expected to occur
IMPROBABLE	E	So unlikely that it can be assumed occurrence may not be experienced	Unlikely to occur, but possible

*Definition of descriptive words may have to be modified based on quantity involved.
†The size of the fleet or inventory should be defined.

More than half of the document's hundred-plus pages consists of a listing of task descriptions. The remainder consists primarily of three appendices.

The task descriptions are broken into three sections. The requirements for system safety programs are to be developed by selecting tasks appropriate for each individual project or program.

MIL-STD-882B CONTENTS

Paragraph 1 Scope
Paragraph 2 Referenced documents
Paragraph 3 Definitions and abbreviations
Paragraph 4 System safety requirements
Paragraph 5 Task descriptions
Task section 100—program management and control
Task section 200—design and evaluation
Task section 300—software hazard analysis

Appendix A   Guidance for implementation of system safety program requirements
Appendix B   System safety program requirements related to life cycle phases
Appendix C   Data requirements for MIL-STD-883B

Task Section 100 lists the program management and control task that could be selected.

TASK SECTION 100:
PROGRAM MANAGEMENT AND CONTROL
100   System safety program
101   System safety program plan
102   Integration/management of associate contractors, subcontractors, and architect and engineering firms
103   System safety program reviews
104   System safety group/system safety working group support
105   Hazard tracking and risk resolution
106   Test and evaluation safety
107   System safety progress summary
108   Qualification of key contractor system safety engineers/managers

Task Section 200 lists design and evaluation tasks.

TASK SECTION 200:
DESIGN AND EVALUATION
201   Preliminary hazard list
202   Preliminary hazard analysis
203   Subsystem hazard analysis
204   System hazard analysis
205   Operating and support hazard analysis
206   Occupational health hazard assessment
207   Safety verification
208   Training
209   Safety assessment
210   Safety compliance assessment
211   Safety review of engineering change proposals and requests for deviation/waiver
212   (Reserved)
213   GFE/GFP system safety analysis

Task Section 300, added by Notice 1 in 1987, lists software hazard analyses.

TASK SECTION 300:
SOFTWARE HAZARD ANALYSIS
301   Software requirements hazard analysis
302   Top-level design hazard analysis
303   Detailed design hazard analysis
304   Code-level software hazard analysis
305   Software safety testing
306   Software/user interface analysis
307   Software change analysis

Each task listed contains three paragraphs. The purpose is the first paragraph, which is followed by the task description and the details to be specified by the MA.

201.1 Purpose. The purpose of Task 201 is to compile a preliminary hazard list (PHL) very early in the system acquisition life cycle to enable the MA to choose any hazardous areas on which to put management emphasis.

201.2 Task Description. The contractor shall examine the system concept shortly after the concept definition effort begins and compile a PHL identifying possible hazards that may be inherent in the design. The contractor shall further investigate selected hazards or hazardous characteristics identified by the PHL as directed by the MA to determine their significance.

201.3 Details to be Specified by the MA (Reference 1.3.2.1).

201.3.1 Details to be specified in the SOW shall include the following, as applicable:

(R) a. Imposition of Tasks 100 and 201.
       b. Format, content, and delivery schedule of any data required.
       c. Identification of special concerns.

An important part of the government system safety effort is to assure that system safety requirements are specified clearly and correctly in contract documents. The formats for system safety products or deliverables are generally prescribed by data item descriptions (DIDs). Other specific requirements concerning the exact documents to be produced and delivered are specified in the contractor requirements data list (CRDL).

Appendix A to the document provides supplemental guidance on almost a paragraph-by-paragraph basis for implementing the tasks and other program elements listed in the body of the standard.

Note that MIL-STD-882B is not the only requirements document in use. For example, MIL-STD-1574 is preferred by some members of the U.S. Air Force Space Division. It serves the same purpose as MIL-STD-882B but was designed to be more "space system specific" (U.S. Air Force 1987). Despite attempts, especially after the issuance of Notice 1 in 1987, to standardize on MIL-STD-882B, MIL-STD-1574 may still be encountered because it was the contractual system safety document on some long-term contracts.

Most major DOD agencies and contractors involved in the traditional weapons and aerospace programs have internal system safety organizations to plan and implement system safety efforts. The majority of the nation's system safety engineers tend to work for these organizations and, by and large, provide the membership for the System Safety Society.

The total system safety effort for complex aerospace programs may involve several levels of organization. A typical six-level system safety effort may be organized as follows:

Level One. Corporate or headquarters. The system safety effort at this level usually consists of general oversight of multiple programs and development of policies and standards.

Level Two. Procurement activity. At the contracting level, the system safety task is to convert requirements and tasks into contractual specifications.

Level Three. Contractor's management system safety program. At this level, contractor management provides program oversight, policy, direction, and resource allocation.

Level Four. Contractor's engineering system safety program. This level is the working level for the system safety engineer and the program system safety effort.

Level Five. Specifications and requirements. Designers and engineers incorporate safety codes, standards, and regulations. This level tends to be compliance oriented.

Level Six. Operational location. At this end user level, users and operators provide user input and review.

An effective system safety working group includes representatives from levels three, four, five, and six (U.S. Air Force 1987).

The DOD system safety programs tend to be more standardized than those of some other agencies, primarily because MIL-STD-882B serves as a foundation for the effort. However, considerable variation in programs, concepts, interpretations, definitions, and approaches occurs, even within the DOD community. This variability is most pronounced in the attempts of contractors to interpret and conform to system safety requirements.

NASA

There are many similarities between NASA's system safety efforts and those of the Department of Defense, especially the USAF Space Division. The organizations tend to share personnel, contractors, regulations, and, to some extent, missions.

The NASA system safety manual. NHB 1700.l(V-3), is one of the several system safety documents supporting NASA system safety efforts.

NASA SYSTEM SAFETY NHB 1700.1(V3)

CHAPTER 1: MANAGEMENT TECHNIQUES

CHAPTER 2: TECHNICAL METHODS

APPENDICES

This particular document, dated 6 March 1970, is one of the older system safety documents in use. The NASA system safety manual was revised and distributed as NHB 1700.1 (V7) (Preliminary) in the mid-1980s.

Some of the other key documents in the NASA system safety effort include

NBH 1700.1(V1A) Chapter 7   Basic safety manual
NHB 5300.4   Safety, reliability, maintainability, and quality
                            provisions for the space shuttle program

NHB 1700.7	Safety policy and requirements for payloads
NSTS 07700 (Volume V)	Information management requirements
NSTS 07700 (Volume X)	Space shuttle flight and ground system specifications
NSTS 22206	Instructions for preparation of failure modes and effects analyses and critical items list
NSTS 22254	Methodology for conduct of NSTS hazard analyses

Commonly used NASA definitions of basic system safety terms include

Hazard: The presence of a potential risk situation caused by an unsafe act or condition (NSTS 22254)

Risk: The chance (qualitative) of loss of personnel capability, loss of system, or damage to or loss of equipment or property (NHB 5300.4[1D-2])

Safety: Freedom from chance of injury or loss of personnel, equipment, or property (NUB 5300.4[1D-2])

System safety: The optimum degree of risk management within the constraints of operational effectiveness, time, and cost attained through the application of management and engineering principles throughout all phases of a program (NHB 5300.4[1D-2])

Hazard analysis types (PEA, SHA, SSHA, OHA) generally parallel those outlined in MIL-STD-882B, even though hazard report formats may vary.

Again the process involves a preliminary hazard analysis to be done very early in the concept stage, followed by subsystem hazard analysis as subsystems are developed, systems hazard analysis that looks at interfaces between subsystems, and, finally, the operating hazard analysis, which tends to add the human element and evaluate procedures.

Common techniques for hazard analysis are the failure modes and effects analysis (FMEA) and fault tree analysis (FTA). Many of the other techniques listed in Chapter 17 are also used. The FMEA is considered a reliability tool and used, in most NASA and NASA contractor organizations, by a separate reliability division or branch. The FMEA is used to generate another popular NASA tool, the critical items list (CIL).

A *critical item* is defined as a single point failure and/or a redundant element in a life- or mission-essential application where:

a. Redundant elements cannot be checked out during the normal ground turnaround sequence.

b. Loss of a redundant element is not readily detectable in flight.

c. All redundant elements can be lost by a single credible cause or event such as contamination or explosion. (NASA 1987)

Even though some NASA documents reference the use of risk assessment codes (RACs), risk assessment for many NASA efforts is based on hazard level or criticality category. If risk assessment codes are used, they tend to use the hazard severity and probability categories and matrices from MIL-STD-882B. The NASA hazard levels are

Catastrophic (CA). No time or means are available for corrective action.

Critical (CR). May be countered by emergency action performed in a timely manner.

Controlled (CN). Has been countered by appropriate design, safety devices, alarm/caution and warning devices, or special automatic/manual procedures. Criticality categories include

1   Loss of life or vehicle
1R Redundant hardware element failure that could cause loss of life or vehicle
1S Potential loss of life or vehicle due to failure of a safety or hazard monitoring system to detect, combat, or operate when required
2   Loss of mission; for Ground Support Equipment (GSE), loss of vehicle system
2R Redundant hardware elements the failure or which could cause loss of mission
3   All others (NASA 1987)

The safety organizational structures and protocol vary widely with the NASA community. At headquarters level, the System Safety Branch is part of the Safety Division. The Safety Division manager reports to the Associate Administrator for Safety, Reliability, Maintainability, and Quality Assurance.

The safety, reliability, and quality assurance (SR&QA) efforts with NASA are complex and organized in such a way that extracting a pure system safety program from the overall SR&QA effort is difficult.

FACILITY SYSTEM SAFETY

Military Construction Programs

Until the mid-1980s, the military safety community typically did not get significantly involved in the facility acquisition cycle until the 90% design review. Occasionally, any real safety participation was actually not until construction was underway or even as late as preoccupancy or acceptance inspections.

During the mid-1980s the U.S. Army Corps of Engineers (USACE) and the Naval Facilities Command (NAVFAC), with input from the other services, initiated efforts to develop facility system safety programs that would provide safety input early in the life cycle of major military construction projects.

Another goal of the effort was to improve end user input into the safety and design effort and overall communications, at all levels, between project management, the engineering community, the safety community, and end users.

The general approach was to tailor MIL-STD-882B requirements to provide a simple, usable, effective approach to facility system safety. Basic definitions, tasks, risk assessment methods, and hazard analyses were straight from the standard.

This facility system safety effort applies to military construction projects managed by USACE and require that the system safety effort begin in the initial stages of the facility acquisition cycle. When a local installation (such as an army post or air force base) identifies the need for a new facility (major military construction project), the acquisition process is initiated by preparing a project development brochure (PDB) and a form 1391 providing the initial project description, justification, and economic analysis. A preliminary hazard list (PHL) is required as part of this initial package, which is sent forward to seek funding. The PHL is prepared by informal conferencing, checklist review, and ETBA. The input required includes analytical trees, flow diagrams, sketches and drawings, an installation map, energy sources, external hazards, and lessons learned. The PHL report consists of a narrative, a list of hazards, and a risk assessment. See Figure 3-1.

PRELIMINARY HAZARD LIST

Project _____
Prepared by _____
Method(s) used: ☐ Informal Conferencing   ☐Checklist Review   ☐ ETBA   ☐ Other _____

Date _____
Page ___ of ___

HAZARDOUS EVENT	CAUSAL FACTORS	SYSTEM EFFECTS	RAC	COMMENTS

Figure 3-1   Preliminary hazard list worksheet.

This PHL is prepared by an installation-level system safety working group (SSWG) consisting of representatives from the engineering, safety, and user communities and project management. A representative from the district Corps of Engineer (COE) office is also recommended. The PHL is normally the only analysis product of this group, but the local SSWG should have an important role in providing input and reviewing the overall system safety effort. This group also influences the level of the system safety effort by categorizing the proposed facility as low, medium, or high risk (Table 3-3).

Low-risk facilities are those with low energy levels and those with which the COE has a considerable amount of trouble-free experience, such as basic administrative buildings and housing. The system safety effort for these facilities may consist primarily of the PHL, with no additional analysis required.

Medium- or high-risk facilities include those that have higher levels of energy associated with them and/or tend to be more unusual because of their function or design. They are given at a minimum a preliminary hazard analysis (PHA) (Fig. 3-2). The results of this PHA determine and drive the need for further hazard analysis in the form of systems hazard analysis, subsystem hazard analysis, and/or operating and support hazard analysis. The technique of choice for facility PHAs is evolving toward energy trace and barrier analysis.

The PHA and any other analyses required are generally performed by the designer, usually a contractor architect or engineer.

The primary responsibility for overseeing the system safety effort rests with the District Corps of Engineer project manager. An SSWG, including the design engineers, safety representative, project management, and representatives from the end user (building occupant) is established at the district level. The higher the risk category, the greater the importance of the input from the end user.

TABLE 3-3. Risk Categorization

Category	Facility Type	Safety Data Required
Low (low user involvement)	Housing, warehouses, administrative buildings	User-prepared PHL only; checklists may be used
Medium (moderate user involvement)	Maintenance facilities, heating plants, photo labs	User-prepared PHL, usually designer-prepared PHA; other analyses may be required
High (heavy user involvement)	Explosive plants, chemical agent facilities, high-energy facilities	User-prepared PHL, usually designer-prepared PHA; other analyses prepared by designer may be required

PRELIMINARY HAZARD ANALYSIS

Project _____

Prepared by _____

Method(s) used: ☐ ETBA ☐ FMEA ☐ FTA ☐ Other _____

Date _____

Page ___ of ___

HAZARDOUS EVENT	CAUSAL FACTORS	SYSTEM EFFECTS	RAC	COMMENTS	RECOMMENDED ACTIONS	CON-TROLLED RAC	CRC	STANDARDS

Figure 3-2 Preliminary hazard analysis worksheet.

The hazard tracking, review requirements, and risk acceptance process are basically the same as for other MIL-STD-882B programs.

Few individuals at the installation level or at COE district level have any system safety experience. Most architect and engineer contractors have no system safety experience. Until more government and contractor personnel are trained, the effort will probably have to rely on the services of specialized facility system safety subcontractors and consultants.

The NAVFAC effort appears to be somewhat decentralized, with efforts being directed primarily from the division level. The NAVFAC approach is also consist with and guided by MIL-STD-882B and generally parallels the efforts of the U.S. Army Corps of Engineers.

THE CHEMICAL INDUSTRY

Another evolving approach to facility system safety is the HazOps effort being developed by the chemical industry. This effort is being promoted by individual chemical and petrochemical companies and by the American Institute of Chemical Engineers.

HazOp is used to describe both a specific analytical tool and the broader, general approach to hazard identification and evaluation in the chemical industry.

The specific tool, the hazard arid operability study, uses a multidisciplinary team to identify, analyze, and control hazards systematically. The HazOp team is virtually the same as a system safety working group in composition and purpose. It traces the flow of materials and products through an operation, concentrating on specific points called *study nodes*. A set of guide words is used to determine types of deviations from prescribed parameters that could occur at each study node. The causes and consequences of the deviations are discussed with the aid of these guide words. Typical guide words and their meanings are

NO	None
LESS	Quantitative decrease
MORE	Quantitative increase
PART OF	Qualitative decrease
AS WELL AS	Qualitative increase
REVERSE	Opposite of intended direction
OTHER THAN	Substitution (Goldwaite 1985)

Operational parameters tend to be energy oriented, such as temperatures, pressures, flow rates, and voltages.

A HazOp study resembles a FMEA in that the guide words and parameters tend to describe failure modes and the consequences examined in HazOp studies parallel the effects described in an FMEA. The HazOp study nodes may have the characteristics of the critical items identified by an FMEA.

The HazOp study is also similar to the ETBA in that it traces energy flows through a facility, studies barriers to control energy flows, and identifies the targets of unwanted energy flows.

The HazOp study differs from the FMEA and ETBA in that some suggest that "the best time to conduct a HazOp is when the design is fairly firm" (Goldwaite 1987). Conventional system safety wisdom dictates that the system safety effort be as far upstream as practical, with a facility preliminary hazard analysis developed as part of the initial design effort and completed by the 35% stage. Also, a HazOp study tends to include human factors and operator errors whereas a traditional FMEA or ETBA normally examines hardware failures only.

Some of the hazard analysis (evaluation) techniques already used by the chemical industry include traditional system safety tools such as preliminary hazard analysis, failure modes and effects analysis, and fault tree analysis.

In summary, the broad HazOp effort of the chemical industry is a facility system safety effort. It is one of the first such efforts to originate in the private sector. A sharing of information between the HazOp advocates and those promoting facility system safety should be mutually beneficial.

DEPARTMENT OF ENERGY

Another major approach tends to be very different from other efforts described. The Department of Energy's approach is based on the management oversight and risk tree (MORT), a comprehensive, state-of-the-art program for safety management and system safety.

The following are definitions of common system safety terms:

Hazard: The potential in an activity (or condition or situation) for sequence(s) of errors, oversights, changes, and stresses to result in an accident

Risk: The probability during a period of activity that a hazard will result in an accident with definable consequences

Safe: A condition wherein risks are as low as practicable and present no significant residual risk

System: An orderly arrangement of components that are interrelated and that act and interact to perform some task in a particular environment.

System safety: Safety analysis (usually specialized and sophisticated) applied as an adjunct to design of an engineered system (Johnson 1980).

Like HazOp, MORT refers to both a specific analytical tool and to a broad approach to system safety.

The program was developed for the Atomic Energy Commission, the forerunner of the Department of Energy, in the early 1970s by Bill Johnson. Johnson developed this program by going out and examining various tools, techniques, and approaches used by other organizations and then developing an approach to system safety tailored specifically for DOE. At the heart of this program is a large analytical or fault tree known as the MORT chart (Fig. 3-3).

Some of the other tools and techniques associated with this program include:

Analytical trees. Graphic trees or diagrams. "Positive" or "objective" trees are used as planning tools, project description documents, and graphic checklists. They are used to help ensure that things go right. "Negative" or "fault" trees are used as troubleshooting or investigative tools to determine what went wrong.

Change analysis. A relatively simple but useful tool that systematically examines changes. Useful as a preventive tool in safety review of proposed changes and as a mishap investigation tool, especially when quick answers are needed or when the specific direct causes of the mishap are obscure.

Event and causal factors charts. A mishap investigation tool that graphically depicts the vents and conditions involved in a mishap and illustrates how

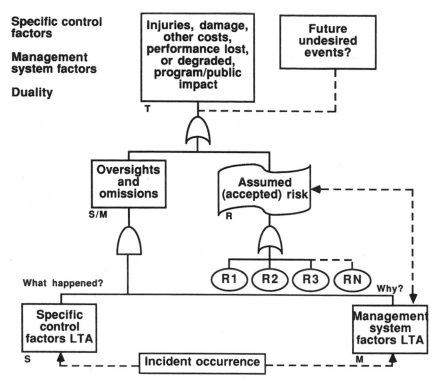

Figure 3-3   The top tiers of the MORT chart. The chart is a fault tree used as both an inspection or appraisal tool and as an accident investigation guide.

they caused or contributed to the mishap. It is an excellent tool to help in planning and organizing an investigation and also an excellent briefing tool and illustration to be included as part of formal accident reports.

Time-loss analysis. A tool developed by the National Transportation Safety Board to evaluate postaccident responses by rescuers, firefighters, medics, and other emergency response personnel.

MORT analysis. A detailed, comprehensive 1,500-item fault tree used primarily for mishap investigation. This valuable tool serves as a graphic guide for organizing and conducting an investigation and for writing the accident report. It may also be used as an appraisal or accident-prevention tool.

Extreme value projection. A relatively quick and simple risk-projection method that aids in determining whether a specific mishap was typical of the operation or a fluke caused by a specific failure or change not typical of the system. Also useful as an appraisal tool, it requires special graph paper.

All of these tools and techniques are an outgrowth of the initial effort by Johnson but had been modified and upgraded by the DOE's System Safety Development Center in Idaho Falls, Idaho. The Department of Energy felt that some degrees of standardization was required for managing the system safety efforts effectively in the nuclear community. It concluded that this standardization could be achieved most readily by developing department-wide approaches and training programs to promote these approaches and by training both DOE (at that time, Atomic Energy Commission) personnel and their contractors.

Even though MIL-STD-882B and MORT tout the importance of a system safety effort throughout the life cycle, both approaches fall somewhat short of full implementation. The MIL-STD-882B efforts cover all phases of the acquisition cycle, but most have no real application after a system has been deployed. In contrast, MORT probably provides the greatest benefits during the operations phase.

The MORT chart contains detailed information to aid in evaluating the hazard analysis process (HAP) and the design effort. In practice, however, the MORT tools and techniques that are most frequently used are in the areas of accident investigation, auditing, appraisal, and performance monitoring.

The MORT tools and techniques can be helpful in preparing a safety analysis report (SAR), the upstream safety product most frequently required for new DOE programs, but the more common system safety products (system safety program plan, preliminary hazard analysis, system/subsystem hazard analysis, operating hazard analysis) are not a dominant part of the MORT program and are seldom even referenced in System Safety Development Center (SSDC) documents.

Thus the MORT approach to system safety tends to be more closely aligned with occupational safety programs. Department of Energy and contractor safety personnel, including those with MORT training and system safety responsibilities, tend to be more closely aligned with the American Society of Safety Engineers (ASSE) and the National Safety Council (NSC) than with the System Safety Society.

A benefit of the DOE approach is that it is standardized and institutionalized within the DOE community.

Accident investigation programs based on MORT are used by NASA and by the ground safety community of the U.S. Air Force.

REVIEW QUESTIONS

1. Briefly discuss the purpose of MIL-STD-882B.
2. Discuss the types of tasks included in Task Section 100 of MIL-STD-882B.
3. Discuss the types of tasks included in Task Section 200 of MIL-STD-882B.
4. Discuss the types of tasks included in Task Section 300 of MIL-STD-882B.

5. What is the purpose of Appendix A of MIL-STD-882B?
6. What are *DIDs* and *CRDLs*? What is the purpose of each?
7. Outline key items of information to be included in an SSPP.
8. What is a *CIL* as defined by NASA? How is a CIL developed?
9. What does NASA define as critical items?
10. Define NASA hazard levels.
11. Define NASA criticality categories.
12. What two government organizations manage major military construction programs?
13. Who would normally prepare a PHL for a facility? When would it be prepared?
14. Discuss the risk categories used in the facility system safety effort.
15. Outline the composition of a typical facility system safety working group.
16. What is *HazOps*? What industry is the primary user of HazOps?
17. Discuss the similarities and differences between HazOps and the MIL-STD-882B approach to system safety.
18. What is *MORT*? What government agency uses MORT as the basis for their system safety effort?
19. List the primary accident investigation tools associated with MORT.
20. List the primary accident prevention tools associated with MORT.

REFERENCES

Goldwaite, William H., and others. 1985. *Guidelines for Hazard Evaluation Procedures.* New York: American Institute of Chemical Engineers.

Johnson, William G. 1973. *MORT, The Management Oversight and Risk Tree.* Washington, D.C.: U.S. Atomic Energy Commission.

Johnson, William G. 1980. *MORT Safety Assurance Systems.* New York: Marcel Dekker.

National Aeronautics and Space Administration. 1970. *NHB 1700.1 (V3): System Safety.* Washington, D.C.: Safety Office, NASA.

National Aeronautics and Space Administration. 1987. *NSTA 22254: Methodology for Conduct of NSTA Hazard Analyses.* Houston: NSTS Program Office, Johnson Space Center.

National Aeronautics and Space Administration. (undated). *Safety, Reliability, and Quality Assurance Phase I & II Training Manual.* Houston: Office of SR&QA, Johnson Space Center.

U.S. Air Force. 1987. *SDP 127-1: System Safety Handbook for the Acquisition Manager.* Los Angeles: HA Space Division/SE.

U.S. Army. (undated). Facility system safety manual. Washington, D.C.: HQ Safety, USACE. Draft.

U.S. Department of Defense. 1984 (updated by Notice 1, 1987). *MIL-STD-882B: System Safety Program Requirements.* Washington, D.C.: Department of Defense.

CHAPTER 4

Problem Areas

Even though the system safety effort is alive and well in the aerospace and nuclear worlds and expanding into other areas, several significant problem areas may be restraining the expansion of new system safety programs and limiting the effectiveness of existing programs.

Eight general problem areas need to be addressed before the system safety effort can provide the safety services that will be needed in the next century. Many of these problems are interrelated.

- Standardization
- Risk assessment codes
- Data
- Communications
- Life cycle
- Education and training
- Human factors
- Software

First, standardization is almost wholly lacking. Each agency, contractor, and analyst has a separate set of definitions, techniques, approaches, and worksheets.

Second, not only is there no *standard* risk assessment code (RAC) but also some people in this field maintain that there is no *adequate* risk assessment code. They are all so subjective as to be almost meaningless.

Third, in some areas reliable, quantitative data are lacking. Meaningful risk assessment must have valid data available.

Fourth, at times some elements of the system safety community have tended to operate out of ivory towers, with narrow concepts and inadequate attention to two of the most valuable sources of information available: end users and the rest of the safety community.

System Safety for the 21st Century: The Updated and Revised Edition of System Safety 2000, by Richard A. Stephans
ISBN 0-471-44454-5   Copyright © 2004 John Wiley & Sons, Inc.

Fifth, even though all system safety programs embrace the life cycle concept, few actually provide meaningful efforts throughout the operations phase. The exception is the Department of Energy's MORT program, which provides less than some programs in the way of upstream analysis but includes an excellent accident investigation program.

Sixth, there is a real shortage of well-qualified system safety engineers. Few engineering schools include safety or system safety courses in their curricula. This problem is greatly compounded by the lack of standardization mentioned earlier. Not only are courses, instructors, and texts lacking but also there is no common body of knowledge to teach!

Seventh, even though some efforts are very good in the area of human factors, most programs could do a better job. Human reliability is extremely difficult to predict accurately. The man-machine interface is an area that deserves more effort.

Finally, there is a general lack of expertise in the area of software system safety. Not only are there grossly insufficient numbers of qualified system safety engineers with software safety credentials but also very few recognized textbooks, seminars, university programs, or competent consultants are available. The lack of technical competence in this area seems to span the entire system safety community.

STANDARDIZATION

One of the cornerstones of the system safety effort to date has been the concept of tailoring, that is, a widely (and deeply) held conviction that program and project managers and system safety analysts should be free to develop a system safety management plan or system safety program plan that is specifically designed for the individual project. This concept was—and is—valid for major research and development programs and many others. The problem is not the concept of tailoring but the lack of standard terms, basic tools, and techniques.

As a matter of fact, a toolbox offers a suitable analogy. Ideally, the system safety manager or engineer has a well-stocked toolbox of analysis types and techniques and is able to study the particular task at hand and select the appropriate tool or combination or tools to perform the task effectively and efficiently. This example is the correct application of the tailoring concept.

However, what if there were no "common" tools? What if each box contained a completely different set of tools in a mixture of standard, metric, and unique sizes and if even the names of the tools were all different? Because no common tools or common names exist, specifying tools—or training mechanics—is almost impossible. Each individual mechanic becomes familiar with certain tools and is generally able to select something that will work, but that competence does not alleviate the problem. Each time we have to change mechanics, we must retrain them. In the real world of system safety, the problems are even greater.

At least two dozen different risk assessment code (RAC) matrices are in use. They all are based on severity on one axis and probability on the other, but the size and layout of individual matrices vary significantly. Some have *severity* on the vertical axis; others have it on the horizontal axis. One makes the low numbers "bad" (an RAC of 1 indicates a high probability of a catastrophic accident), whereas the next uses a 1 to indicate a very remote probability of an insignificant loss. Some use RACs from 1 to 4; others use 1 to 16; still others use a combination of numbers, roman numerals, and/or letters (1I to 4V or 1A to 5D).

The prioritized list of controls for hazards may be called the *safety precedence sequence*, the *hazard reduction precedence*, the *hazard control sequence*, the *system safety sequence*, or any of several other terms. The only real difference between an operating hazard analysis (OHA) and an operating and support hazard analysis (O&SHA) is the name.

The event and causal factors charts used by the Department of Energy as an accident (or mishap) investigation tool is basically the same tool as multilinear event sequencing (MES) and similar to simultaneous timed event programming (STEP)—all based on concepts of Ludwig Benner. It is also called *causal factors analysis* (CFA).

Laundry lists of analyses frequently mix types of analyses (preliminary hazard analysis, system hazard analysis, and operating hazard analysis) with the methods or techniques for performing analyses (fault tree analysis, energy trace and barrier analysis, failure modes and effects analysis, common cause analysis, change analysis, and so on). Whether *fault hazard analysis* is a type or a method depends upon the reference in use. For all practical purposes, fault hazard analysis and system (or subsystem) hazard analysis seem to be the same thing, which is apparently called *gross hazard analysis* occasionally.

The first two pages of the U.S. Air Force's SDP 127-1, *System Safety Handbook for the Acquisition Manager*, states:

> It is difficult to explain the "whys" and "hows" of the System Safety discipline when there is a lack of agreement within the discipline as to just what the task really is. At a meeting of approximately 50 System Safety engineers, each engineer was asked to provide a definition of system safety. Of those 50 fully-qualified and experienced System Safety engineers, at least 30 had distinctly different ideas of what constitutes the system safety task. Very little standardization currently exists between agencies or even between the directives, regulations, and standards that implement the requirement. (USAF 1987)

The lack of commonality makes extremely difficult the task of training system safety personnel, evaluating and comparing programs, and effectively monitoring and controlling system safety efforts. The barriers to effective communications resulting from the different languages make the process of contracting for system safety services a sometimes confusing or risky business for contractors and clients.

An unconfirmed story about one of the military services has it advertising for a series of system safety analyses. The prices bid for the work ranged from around $30,000 to over $800,000! This disparity reflects the differences in interpretation of system safety requirements throughout parts of the system safety community. (The contract was reportedly awarded to the low bidder.)

At some point the proliferation of techniques, worksheets, definitions, formats, and approaches has to end, or at least some common ground has to be established.

The model program in Part 2 and other portions of this book can provide some of that common ground.

RISK ASSESSMENT CODES

The problems with risk assessment codes, unfortunately, go beyond the lack of standardization. Even though a great deal of benefit can be derived just from the interaction of a system safety working group and the design improvements its members generate, the system safety effort should contribute the accurate, meaningful analysis of hazards, and should produce objective, useful data upon which enlightened decisions can be made. The bottom line of the analysis effort is, for many programs, the risk assessment code. The RAC should provide a valid basis for determining the acceptability of risks, prioritizing risks, and allocating resources to reduce risks. The quality of the RAC, then, to a great extent determines the overall value and credibility of the system safety effort.

Unfortunately, most RAC matrices use scales that are so subjective and poorly defined as to be virtually meaningless. Granted that accurately placing exact numerical values on either severity or probability scales is difficult, these scales must be quantified to some extent to have any real value. The "hurts a little—hurts a lot" and "maybe—maybe not" approach to severity and probability scales, respectively, hardly provides the type of data required to support major risk acceptability decisions.

Additionally, the severity scales are too restrictive. Certainly the loss of a single human life is tragic: if the loss of a single life represents the upper extreme on the severity scale, however, how are hazards having the potential for killing dozens or hundreds or thousands of people in a single mishap to be rated?

Similarly, the probability scales are, in many instances, either broadly defined subjective word descriptions or, if quantitative, calculated guesses with no supporting calculations. In order to produce probability numbers, assumptions may frequently be necessary about failure rates, the number of units in the system, and the expected service life of the unit or the system. These assumptions must be clearly stated and identified so that RACs can be updated, revised, or validated as historical or test data are developed. Frequently, determining how the probability "guess" was determined is difficult.

Both severity and probability scales need to be expanded to provide more meaningful RACs. With many RAC matrices in use today, the opinions of ten different experts tend to produce ten different RACs. Reducing the reported level of risk by seeking a second opinion on the RAC associated with the risk becomes extremely tempting.

If RACs are, in fact, going to serve as the drivers in the system and the basis for risk acceptance and resource allocation decisions, the numbers generated must have a positive correlation to the risk involved and to the resources required to control the hazard. A universal RAC matrix would provide a means of comparing relative risks associated with multiple projects, evaluating the allocation of funds, and determining the cost-effectiveness of various controls. Ideally, this universal risk assessment code matrix could provide an evaluation of risks in absolute as well as relative terms. Chapter 12 contains a proposed universal RAC matrix (Total Risk Exposure Codes). The use of a meaningful, quantified, expanded universal RAC matrix may represent a practical approach to probabilistic risk assessment.

DATA

In order to produce believable RACs or any other quantitative risk assessment, reliable, valid data are required. Even though considerable data exist, they are not necessarily available or in the correct format. Improvements can be made in the sharing of lessons learned, mishap information, reliability data, and the other information needed to support the system safety effort. A well-organized effort to identify and catalog existing databases and to develop plans for the systematic collection and dissemination of new data would benefit the entire safety community.

COMMUNICATIONS

The fourth problem is not nearly as important or serious as the first three. As a matter of fact, lack of communication and loss of information are age-old concerns for most organizations. In the system safety effort these concerns are sometimes the result of system safety, acquisition, and design personnel who are physically and/or organizationally remote from the end users and the local safety organization. The system safety effort requires that system safety engineers, managers, and designers seek input from local safety organizations and from end users. Equally important, field personnel must aggressively push upstream lessons-learned information, accurate project requirements, and other real world input. A well-organized system safety working group approach eliminates this communications problem. Unfortunately, all the right people are not always included in groups.

A well-organized SSWG can also be used to overcome another apparent shortcoming of some existing system safety programs.

LIFE CYCLE

Although all approaches to system safety give considerable lip service to life cycle or "cradle-to-grave" or "womb-to-tomb" programs, in practice many programs based on MIL-STD-882B concentrate their efforts on preoperation phases. Once a system is in full production and fielded, very little additional system safety effort is made.

In contrast, the Department of Energy's MORT-based approach provides excellent accident investigation, audit, and appraisal tools oriented to ongoing operations. Despite emphasis on the importance of the upstream process, however, system safety programs based on MORT tend to lack the early, detailed, systematic hazard identification and analysis efforts that are characteristic of MIL-STD-882B programs.

Late life cycle system safety efforts may draw more attention as some of the major systems age. A classic example is the commercial aviation fleet. As expensive aircraft approach the end of their projected service life, new system safety studies may be required. Other significant late life cycle system safety efforts may be required to address the demilitarization of chemical munitions, disposal of nuclear weapons, and the aging of major structures. The approach suggested in Part 2 addresses life cycle system safety.

EDUCATION AND TRAINING

A sixth problem confronting the system safety effort is the lack of qualified system safety engineers and managers, even for system safety efforts in place at the beginning of the 1990s. If the system safety effort is to expand to meet the challenges of the next century, many more personnel will be required. Additionally, they will all need to know system safety objectives, concepts, and methods in order to participate in SSWGs and to interface with the overall effort.

Currently, system safety education and training needs are met by in-house programs, seminars, and workshops offered by consultants, private and professional organizations, and one or two major universities. Even these programs target a relatively narrow audience, typically those already assigned to a "system safety" position! To meet the needs of the twenty-first century every major engineering program in the country needs at least one system safety course in its core curriculum. Advanced courses and courses for managers and safety professionals must also be available. The lack of standardization previously discussed is a key factor in limiting the development of system safety training.

HUMAN FACTORS

Meaningful analysis of the people part of systems tend to be a problem for some parts of the system safety community for two reasons.

First, humans have many more failure modes and are far less predictable than hardware. Accurately determining human reliability is extremely difficult. Considerably less reliable data are available on human performance than on hardware.

Second, few engineering programs prepare graduates for evaluating human behavior and human performance. Chapter 8 provides an overview of basic human factors engineering.

SOFTWARE

Finally, the area of software system safety lacks not only qualified practitioners and education and training programs but also apparently knowledge. Computers and artificial intelligence are playing an increasing role in the control of everything from major weapons systems to home appliances, and the need in this area of system safety is already great. Given the current lack of capability in this area and the almost certain rapid growth, clearly the need for research and development efforts is urgent. Computer-controlled systems going amok may be one of the most significant hazards of the twenty-first century.

REVIEW QUESTIONS

1. List the eight problems associated with the system safety effort.
2. Which problems can be significantly ameliorated by the use of well-organized and well-managed system safety working groups? Explain.
3. Explain how the lack of commonality of system safety terms and approaches affects the education and training of system safety engineers and managers.
4. In what area of system safety is the gap between needs and the availability of qualified practitioners growing fastest?
5. Discuss briefly the shortcomings of the risk assessment codes currently in use.

REFERENCE

U.S. Air Force, 1987. *System Safety Handbook for the Acquisition Manager.* SDP 127-I. Los Angeles: HQ Space Division/SE.

CHAPTER 5

The Future of System Safety

MORE FIRST-TIME SAFE SYSTEMS

System safety evolved as a discipline because of a compelling need for first-time safety in systems where accidents could be catastrophic. During the three decades since the origins of system safety, the number of systems with potential for catastrophic failure has dramatically increased. In addition to expanded requirements in the nuclear and aerospace industries, catastrophic losses in terms of injury, property damage, and environmental impact can result from single mishaps in surface transportation, chemical operations, facilities, and conventional weapon systems. Mass production and distribution of flawed products and services can also produce disastrous losses. The original motivation for system safety is still present but now applies to more and more areas.

As technology advances, everything tends to get bigger and faster. Higher energy levels are required, and more targets are available for unintentional energy flows. Also, as the rate of change in society increases, the gap between compliance safety (provided by codes, standards, and regulations generated by traditional trial-and-error experience) and optimum safety (provided by system safety) tends to grow wider.

Additionally, increased competition in the world market will tend to demand greater speed of production and shorter upstream efforts. Survival in this competitive environment will depend upon an organization's ability to "do it right the first time, every time."

COST-EFFECTIVE MANAGEMENT TOOLS

The greatest driver in the system safety effort as we move into the twenty-first century will be the realization that a well-managed system safety program is cost-effective. It contributes to better design and better products—not only better in terms of safety but also better in terms of customer satisfaction.

System Safety for the 21st Century: The Updated and Revised Edition of System Safety 2000, by Richard A. Stephans
ISBN 0-471-44454-5   Copyright © 2004 John Wiley & Sons, Inc.

Most system safety tools are also management tools. System safety working groups with representation from key organizational units and end users are very similar in concept and operation to quality circles and other methods of increasing involvement and participation and improving communications throughout the organization. Analytical trees are not only excellent project description documents and feeder documents for system safety analyses, but also outstanding planning tools, status charts, and feeder documents for program evaluation and review technique (PERT) or critical path method (CPM) charts. Failure modes and effects analyses provide information that can aid in the safety effort but are primarily reliability tools. A knowledge of human factors improves product safety, industrial safety, production, employee relations, customer relations, and marketing.

THE NEW FACE OF SYSTEM SAFETY

The continuing and growing need for first-time safe systems will require system safety efforts for more and more products and services, in government and in private industry. These efforts must include

- Standardized methods
- Universal RAC
- Late life cycle efforts
- Team effort
- Private industry programs
- System safety courses
- Improved data and communications
- Emphasis on human factors
- Emphasis on software

Even though tailoring will continue to be an important system safety concept, standardized tools and techniques will evolve. System safety will become more of a science and less of an art, and the quality of the systems safety products will be determined more by adherence to accepted procedures than by the individual skill of the analyst. The price for system safety services will come down and the quality will go up. There will be no place (and never has been) for "paper" programs. The risk assessment information generated by hazard analyses will tend to be more accurate and quantitative with the use of better data bases, more meaningful risk assessment codes, and standardized techniques. More late life-cycle system safety efforts will emerge, concentrating on accident analysis, operation of systems toward the end of projected service life, and disposal. System safety courses will become a part of the core curriculum in many engineering, safety, and management programs. Research efforts will improve capabilities in human factors and in software safety analysis.

All of this will happen because it has to happen.

Software System Safety

As an overview of software system safety, the following is provided from the website of the U.S. Army Communications-Electronics Command (www.monmouth.army.mil/cencom/safety/main.htm). "Software system safety optimizes system safety in the design, development, use, and maintenance of software systems and their integration with safety critical hardware systems in an operational environment."

Software system safety is an element of the total safety and software development program. Software cannot be allowed to function independently of the total effort.

A software specification error, design flaw, or the lack of generic safety-critical requirements can contribute to or cause a system failure or erroneous human decision. To achieve an acceptable level of safety for software used in critical applications, software system safety engineering must be given primary emphasis early in the requirements definition and system conceptual design process. As shown in Figure 5-1, safety-critical software must then receive continuous management emphasis and engineering analysis throughout the development and operational life cycles of the system.

Design Criteria

Design for Minimum Risk. Eliminate identified hazards or reduce associated risk through design. Greater complexity in the system increases the possibility that design faults will occur and emerge into the final product. The complexity of safety-related software should be measured and be under management control. Interface design should emphasize user safety instead of user friendliness. User-friendly functions in software can increase the level of risk. To design operator interfaces for safety critical operations, human factors should be considered to minimize the risk created by human error. The system must be designed to be testable during development. Priorities and responses must be analyzed such that the more critical the risk, the higher the response priority in the software.

Tolerate the Hazard. The design needs to be fault tolerant. That means, in the presence of a hardware/software fault, the software still provides continuous correct execution. Consider hazard conditions to software logic created by equipment wear and tear, or unexpected failures. Consider alternate approaches to minimize risk from hazards that cannot be eliminated. Such approaches include interlocks, redundancy, fail-safe design, system protection, and procedures.

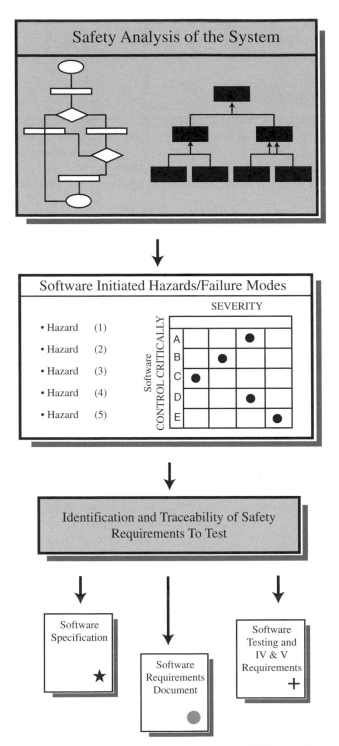

Figure 5-1   System safety process that includes software. (U.S. Army Communications-Electronic Command, 2003).

Fail-Safe Degradation. The software should be designed to limit failures' effects on normal operation. This kind of containment prevents the development of hazardous conditions. Hardware and software faults should be detected by the software, and effective fail-safe exits should then be designed into the system. When applicable, provisions should be made for periodic functional checks of safety devices. This results in needs for start-up built-in-test (BIT), continuous BIT, redundancy check, and other design approaches intended to help ensure the correct functioning of critical components and its handling in a degraded mode should a failure occur. The degraded operation mode must be well thought out, since it encompasses many interacting system components.

Provide Warning Devices. When neither software design nor safety devices can effectively eliminate identified hazards or adequately reduce associated risk, devices should be used to detect the condition and to produce an adequate warning signal to alert personnel of the hazard.

Develop Procedures and Training. Not all hazards can be controlled within the software subsystem. Providing adequate hazard controls is a system level issue that includes the control of physical hazards from all areas, and minimizing risk resulting from environmental conditions. Where it is impractical to eliminate hazards through design, procedures and training must be used. All software-related functions that affect service should be documented, and users should be fully trained.

Specific Development Activities. Software life cycle activities related to system safety engineering include the following:

- Preliminary hazard analysis (PHA)
- Software requirement hazard analysis (SRHA)
- Design specification hazard analysis (DSHA)
- Subsystem hazard analysis (SSHA)
- System hazard analysis (SHA)

PROACTIVE OR REACTIVE?

The only question is whether the effort will be proactive or reactive. Proactive efforts are led by the enlightened leadership of safety, engineering, and management communities; reactive efforts are pushed by accidents, litigation, and overseas competition.

With the proper leadership and initiative from members of the existing safety community, supported and aided by thoughtful engineers and managers, this effort can aid in providing safe products and services and in gaining a competitive edge as we move into the next century.

Without enlightened and aggressive leadership, the necessary expansion and improvement in the system safety effort may come as the reaction to catastrophic accidents, litigation, or initiatives begun in other parts of the world.

System safety may grow as a separate discipline or the system safety effort may be absorbed into the mainstream of industrial safety, loss prevention, risk management, loss control, or some other program. A new name or buzz-word may appear. Nevertheless, the need for first-time safe systems and for the application of system safety principles, tools, and techniques to systematically identify, analyze, and control hazards as early in the life cycle as possible (with continuing efforts throughout the life cycle) will continue to grow indefinitely.

REVIEW QUESTIONS

1. What factors will drive expansion of the system safety effort in the future?
2. What will be the most important motivation for future system safety efforts? Explain.
3. Discuss the benefits of a system safety effort other than improved safety.
4. Discuss how the system safety efforts of the future will differ from existing programs.
5. Explain the gap between compliance safety and optimum safety, and discuss its significance to future safety efforts.

REFERENCE

U.S. Army Communications-Electronics Command Website. Available at www.monmonth.army.mil/cencom/safety/main.htm. Accessed June 19, 2003.

PART II

SYSTEM SAFETY PROGRAM PLANNING AND MANAGEMENT

CHAPTER 6

Establishing the Groundwork

The purpose of this section is to propose a systematic method of establishing (or upgrading) a system safety program for the following target audience:

- Public and private organizations with no system safety experience and no externally imposed system safety requirements
- New government contractors and subcontractors who are required to develop and implement system safety programs to meet government requirements but have limited system safety experience
- Government and prime contractor safety organizations with oversight and review responsibilities for system safety programs.

GENERIC MODEL

The generic model program outlined here is generally consistent with MIL-STD-882B requirements but contains additional elements and concepts. For example, the generic model program attempts to address all phases of the life cycle of a system to a greater extent than does MIL-STD-882B, which concentrates primarily on all phases of the acquisition cycle.

Also, the generic model program is designed primarily for application to relatively simple systems. The traditional system safety community has had the responsibility for establishing effective system safety programs for aerospace, nuclear, and advanced weapons systems. The generic model is not as sophisticated or complex as the programs conducted by major government contractors such as Boeing, McDonnell Douglas, Lockheed, General Dynamics, and Rockwell. The model program presented is intended to aid in expanding the system safety effort into such areas as facilities, surface transportation, utilities, and general industry. Thus, the systems to be made safe are generally less complex and involve lower energy levels and fewer hazards than the systems originally analyzed by the system safety effort.

System Safety for the 21st Century: The Updated and Revised Edition of System Safety 2000, by Richard A. Stephans
ISBN 0-471-44454-5   Copyright © 2004 John Wiley & Sons, Inc.

PRODUCT SAFETY

At this point, a discussion of the relationship, as used in this book, between system safety and product safety is appropriate.

As defined in Part 1, a *system* is a composite consisting of people, procedures, and facilities and/or hardware working in a given environment to perform specific tasks. A *product* is generally just the hardware part of the system. Thus, if we consider not just a screwdriver, but the screwdriver, the people who use the screwdriver, the procedures or instructions needed to use the screwdriver, the environment within which the screwdriver is used, and the tasks for which the screwdriver is used, the screwdriver becomes a "screw installation and removal system."

The distinction between the generic model and others becomes, then, the inclusion of personnel, procedures, environment, and task considerations, not the complexity of the hardware. Whereas product safety may include only hardware, system safety includes hardware, personnel, procedures, environment, and task considerations, and a safety system approach is appropriate as a method of achieving product safety regardless of the complexity of the hardware. The approach to system safety outlined in the following chapters is also a state-of-the-art method of achieving product safety.

To date, all components of the system have been considered in system safety programs, but the focus has been primarily on a facility or a piece of hardware such as an aircraft, missile, or bomb.

The generic model suggests that the same methodology may be appropriate for the design of toasters, tankers, buses, houses, tools, service stations, and, indeed, virtually any product.

DUAL PROGRAMS

As a matter of fact, the system safety effort for a widget factory would include a program to identify, analyze, and control hazards associated with widgets and another program to identify, analyze, and control the hazards associated with the factory and the manufacturing process necessary to produce the widgets.

A system safety effort would also be appropriate for a hospital, hotel, restaurant, travel agency, or department store. A systematic, upstream effort to identify, analyze, and control hazards is an appropriate way to improve workplace safety even in service organizations. Again, two programs may be appropriate, one that focuses internally on the safety of employees and staff and a second that focuses on the clientele.

PLANNING AND DEVELOPMENT METHODOLOGY

The foundation for the generic model is based on some areas of common ground and some attempts to bridge gaps between individual approaches now in use. This model attempts to take the best from existing approaches and add new ideas to produce some basic planning. Steps toward developing a comprehensive, cost-effective system safety program

1. *Determine where you want to go.* Examine the generic model. Modify it as necessary to meet the needs of your organization. (Assuming this model is used as the basis for teaching system safety, keeping modifications to a minimum will be to your advantage.)
2. *Determine where you are.* Compare your existing system safety program with the model (very easy if you do not have a program!) and systematically list all differences.
3. *Develop a detailed plan to close the gap.* Tailor and edit the procedures given here, if necessary, to produce a systematic plan for transforming your present program into the ideal (model) program.

REVIEW QUESTIONS

1. Discuss the difference between a product and a system.
2. Discuss the concept of dual system safety programs. What is the focus of each?
3. Is a system safety effort appropriate for service organizations? Explain.
4. List the three general steps involved in initiating the planning and development of a system safety program.

CHAPTER 7

Tasks

The tasks required to establish, conduct, and maintain an ongoing system safety effort fall into three broad categories: planning tasks needed to initiate the program, primary system safety tasks (identifying, analyzing, and controlling hazards) needed to conduct the program, and support tasks needed to maintain the program.

Planning tasks are the responsibility of line management and are normally performed by a system safety planning group, supported by the safety staff, and reviewed and approved by corporate and project management. These tasks generally involve developing requirements, policy, and criteria and are discussed in detail in Chapter 9.

The primary system safety tasks are hazard identification, hazard analysis, and hazard control. Primary system safety tasks are normally prepared by an SSWG consisting of representatives from management, safety, engineering (the design organization), and the end user organization or group.

Support activities such as training, documentation, and data collection are normally performed by the safety staff.

These tasks must be performed throughout the life cycle of any project. The phases of the life cycle are as follows (Fig. 7-1):

Concept phase: The initial phase of any project, during which general objectives and specifications are developed.

Design phase: General objectives and specifications are made specific.

Production phase: The plans and specifications are converted into the end product.

Operations phase: The end product is in use during this phase.

Disposal phase: The end product is discontinued, destroyed, or otherwise discarded.

System Safety for the 21st Century: The Updated and Revised Edition of System Safety 2000, by Richard A. Stephans
ISBN 0-471-44454-5   Copyright © 2004 John Wiley & Sons, Inc.

CONCEPT PHASE - General objectives, descriptions, and requirements are developed

DESIGN PHASE - Specific plans, drawings, and specifications are developed

PRODUCTION PHASE - End product is manufactured/constructed/fabricated

OPERATIONS PHASE - End product is in use

DISPOSAL PHASE - End product is destroyed/discarded

Figure 7-1   Life Cycle Phases.

The end product can be a facility, an item of hardware, or even a service, regardless of complexity. Depending on the nature and complexity of the end product, any or all of these stages may be subdivided.

For example, the concept phase may involve an initial concept phase and a definition phase. The design phase may include initial design and final design stages or be divided into distinct segments by percentage of completion, such as 10% design, 35% design, 60% design, 90% design, and final design. These segments may be distinct review points or subphases of facility design.

The production phase may include development and testing iterations before full production or even development, testing, and limited production. The production stage may be called the *fabrication* or *construction* phase. Separate installation, testing, and/or acceptance phases may be inserted after limited and/or full production.

The operations phase should probably be subdivided into early, middle, and late subphases for most products and for most system safety efforts.

The disposal phase may include disposal planning, phase-out, tear-down, or destruction and/or postdisposal or closeout phases.

HAZARD IDENTIFICATION

Hazard identification includes determining what parts of the project constitute a hazard and determining the location of these hazards. Hazard identification continues throughout the life cycle but is concentrated in the concept and design phases and when changes are made or accidents occur (Fig. 7-2).

Concept Phase

During the concept phase, hazard identification is initiated and documented by preparing a preliminary hazard list (PHL).

CONCEPT PHASE - PHL

DESIGN PHASE - PHA, SSHA,SHA, OHA

PRODUCTION PHASE - OHA, Change Analysis

OPERATIONS PHASE - Change Analysis, JSA,
Accident Analysis, Inspections

DISPOSAL PHASE - OHA, Change Analysis, JSA

Figure 7-2   Hazard Identification.

The methods used for preparation of a PHL include informal conferencing and the use of checklists and/or energy trace and barrier analysis (ETBA).

Informal conferencing includes studying the initial drawings, sketches, or plans for the end product, determining the energy sources and materials involved, examining the characteristics of the personnel who will be using the end product, evaluating the location and/or environment in which the end product will be used, and considering the tasks for which the end product may be used.

Input data for preparation of the PHL include project description documents (narrative descriptions, sketches, preliminary drawings, artists' concepts, drawings or photographs of existing similar products, and analytical trees), historical data (safety, reliability, and loss data and accident reports from similar end products and/or components), and any other relevant studies or research.

Hazard discovery forms are useful feeder documents for preliminary hazard lists.

Design Phase

Hazard identification is continued throughout the design stage and documented in the preliminary hazard analysis (PHA), subsystem hazard analysis (SSHA), and the system hazard analysis (SHA). Even though the primary purpose of these products is to analyze previously identified hazards and to determine the adequacy of controls, every effort should be made to continue to identify new hazards, especially those associated with interfaces and changes.

In addition to the techniques used during the concept phase, change analysis should be used to identify any new hazards resulting from changes made during the design phase.

Production Phase

Hardware hazard identification during the production phase is normally limited to complex projects requiring limited production and testing subphases and to other changes that may be made during fabrication, construction, or installation. Change analysis is the preferred technique to identify hazards resulting from changes during this phase.

The operating hazard analysis (OHA) is normally started and sometimes completed during this phase, even though, if adequate information is available, the OHA may be initiated in the design phase. Hazards associated with the human interface and with operating and maintenance procedures should be identified at this time. Project evaluation tree (PET) analysis is the preferred technique for performing OHAs.

Additionally, traditional occupational and industrial safety methods should be employed to identify and control construction and manufacturing hazards associated with the project. For the remainder of the life cycle, system safety efforts are complemented by existing occupational, industrial, or operations safety programs.

Operations Phase

During the operations phase, new hazards are identified by periodic inspections, worksite monitoring, audits, and appraisals. Techniques used to identify hazards during the operations phase include checklists, PET analysis, and safety studies. Accident analysis is also an important method of detecting previously undetected or uncorrected hazards. Accident analysis tools include change analysis, PET analysis, MORT and mini-MORT analysis, and event and causal factors charts.

Again, traditional occupational and industrial safety tools like job safety analysis (JSA) should also be used.

Disposal Phase

An operating hazard analysis and/or a job safety analysis should be performed to identify hazards associated with disposal operations.

HAZARD ANALYSIS AND CONTROL

Hazard analysis involves evaluating each identified hazard by determining the potential severity of accidents resulting from each hazard and the probability that such an accident will occur.

Hazard control involves applying the safety precedence sequence to identified hazards that present unacceptable risks until the risks are controlled to an acceptable level. Inherent in the analysis and control process is verification

CONCEPT ' DESIGN 'PRODUCTION'   OPERATIONS   ' DISPOSAL

CONCEPT PHASE - (Normally very little)

DESIGN PHASE - PHA, SSHA,SHA, OHA

PRODUCTION PHASE - OHA, Change Analysis

OPERATIONS PHASE - OHA, Change Analysis, JSA,
        Accident Analysis

DISPOSAL PHASE - OHA, Change Analysis, JSA

Figure 7-3   Hazard Analysis and Control. Like hazard identification, hazard analysis and control must be life cycle efforts.

that applicable codes, standards, and regulations have been met. The process through the phases of a project is shown in Figure 7-3.

First, every attempt is made to alter the design to eliminate the hazard or to reduce it to an acceptable level. If that is not practical, physical guards and barriers are considered to separate potential unwanted energy releases or other hazards from potential targets. After physical guards and barriers have been established to the greatest extent practical, warning devices are installed, if practical, to provide sufficient warning to shut down the system, neutralize the potential hazard, and/or evacuate the area. Finally, after all other control measures have been considered, procedures, training programs, and other administrative controls are put in place to control further any unacceptable risks. The residual risks that remain after these controls are in place are then formally accepted by management in accord with program criteria. If the risks have not been controlled to an acceptable level, new efforts at control must be made, management must alter or make an exception to established criteria, or the project must be abandoned.

Concept Phase

Very little real analysis is completed during the concept phase because analysis detail and data are generally not available. A preliminary risk assessment code (RAC) is determined, however, as part of the preliminary hazard list. This initial RAC is used to aid in determining the initial scope of the system safety effort and in the early evaluation of alternative designs and approaches.

Even though no formal attempt is made to control hazards and controls are normally not addressed in preliminary hazard lists, some control may be achieved by eliminating obviously hazardous approaches during the initial screening of concepts.

Design Phase

The major hazard analysis and control effort is concentrated in the design phase (or phases). The first and in many ways the most important hazard analysis is the preliminary hazard analysis (PHA).

The PHA is vital because it is the first hazard analysis performed. Ideally the PHA should be conducted in conjunction with the initial design and be completed very early in the design process, when making changes is easiest. The PHA report provides the basis for determining the level of effort for further hazard analyses.

After the PHA is complete, first subsystem hazard analysis (SSHA) and, if required, system hazard analysis (SHA) are performed. Depending on the nature and complexity of the end product and the results of the PHA, SSHAs may be performed on all subsystems or just on selected critical subsystems. Unlike MIL-STD-882B, software analyses are not generally identified separately. If applicable, preliminary software hazard analysis is part of the PHA. Software should be treated as a subsystem and, if further software analysis is required, an SSHA can be performed on the software.

The recommended techniques for preliminary hazard analysis are energy trace and barrier analysis (ETBA) and failure modes and effects analysis (FMEA). Recommended techniques for system and subsystem hazard analyses are FMEA, fault tree analysis (FTA), common cause analysis, sneak circuit analysis (for electrical, electronic, and some hydraulic or pneumatic circuits) and, of course, software hazard analysis for software.

Production Phase

In theory, if adequate analysis and control tasks were completed during design, no additional effort is required on the end product during the production phase. For complex, high-hazard, or high-technology end products, however, the production phase may include the fabrication and testing of prototypes and resulting design changes. In this case, change analyses should be conducted to ensure that changes do not introduce unacceptable risks. If significant changes are made, additional subsystem or system hazard analyses may be necessary.

In all cases, a very important analysis and control effort should be initiated and, if sufficient data are available, completed during the production phase. The operating hazard analysis (OHA) focuses on the human interface with the end product. It examines the adequacy of maintenance and operating procedures and instructions and, if appropriate, the adequacy of the organization and training of maintenance and operations personnel. The recommended technique for completing the OHA is the project evaluation tree (PET) analysis.

Operations Phase

During the operations phase, the analysis and control of hazards identified during traditional inspections and worksite monitoring are ongoing.

At periodic intervals during the life cycle, OHAs are performed. The frequency is determined by the nature of the system, the projected life of the system, and the actual accident experience. Typically, the frequency of inspections and hazard analysis and control efforts is high during the early life cycle, lowest during the middle, and increasing as the end product approaches the end of its projected life.

Other important hazard analysis and control efforts during the operating phases are accident analysis and change analysis. Hazards identified as the result of accidents and changes must be systematically analyzed and controlled to an acceptable level of risk in the same manner as those identified during design.

Project evaluation tree (PET) analysis is recommended as the primary evaluation technique.

Disposal Phase

The analysis and control of hazards during the disposal phase should be addressed with an operating hazard analysis, a job safety analysis (a occupational and industrial safety tool), and/or change analysis.

SYSTEM SAFETY SUPPORT TASKS

System safety support tasks are those tasks, normally performed by the safety staff, required to support and maintain system safety efforts, including verification and tracking, training, contracting support, and data collection and retrieval.

Verification and Tracking

A systematic method must be established to document and track each identified hazard throughout the life cycle and to ensure that control measures are, in fact, implemented. The method recommended is to use a hazard tracking log as a permanent part of the project documentation.

Training

To implement a system safety program, the personnel involved have to be properly trained. A comprehensive training program is particularly important until system safety courses are established in the curricula of engineering, management, and safety majors in colleges and universities. A recommended training plan is included in Chapter 9.

Contracting Support

A contracting effort may become an element in a system safety program in a number of ways. Tradition a1 system safety efforts have virtually all involved a procurement or acquisition process, by the government, of a major end product. The government outlined system safety requirements and the contractor(s) developed and implemented system safety programs to meet those requirements. As system safety programs expand into other government agencies and into private industry, more contracts can be expected to contain system safety requirements. Additionally, even in-house programs may find contracting for system safety training and/or analysis services to be necessary and/or cost-effective. The safety organization may be involved in aiding procurement personnel with developing statements of work and requests for proposals for system safety efforts. Suggestions for contracting language are given in Chapter 9. The safety organization may also serve as technical evaluator and/or technical contact for contracted system safety services.

Data Collection and Retrieval

A data collection and retrieval system must be developed to support the system safety effort. This system includes project description and historical data as well as hazard tracking information.

REVIEW QUESTIONS

1. What are the three broad categories of system safety tasks? Who is generally responsible for each category?
2. What are the primary system safety tasks?
3. What are the primary life cycle phases?
4. What types of analyses are generally conducted during the design phase?
5. What phase would include the production and testing of prototypes?
6. What are the three recommended subphases of the operations phase?
7. During which of the operations subphases would the system safety level of effort be expected to be lowest? Why?
8. Name the system safety support tasks. Who is normally responsible for conducting or coordinating these tasks?
9. Discuss how traditional occupational and industrial safety inspections and worksite monitoring programs interface with the system safety effort.
10. Discuss situations in which a contract may be required as an element of a system safety effort.

CHAPTER 8

System Safety Products

Most of the system safety effort involves providing a service. That service is to identify, analyze, and control hazards as early in the life cycle as possible in order to produce cost-effectively a safer end product. Several products are produced as part of the system safety effort. These products (all documents) communicate and document risk information to management and provide a means of monitoring and auditing the effort.

SYSTEM SAFETY PROGRAM PLAN

Purpose

The purpose of the system safety program plan (SSPP) is to identify the specific system safety requirements for a given project, to include specified system safety tasks, risk assessment methodology and risk acceptance criteria, system safety products and milestones, and system safety organization.

Time of Preparation

The SSPP is prepared at the initiation of the project and updated at each review point, generally at the end of each life cycle phase.

Preparer

The SSPP is prepared by the system safety manager and the safety staff and is reviewed and approved by the general manager and the project manager.

System Safety for the 21ˢᵗ Century: The Updated and Revised Edition of System Safety 2000,
by Richard A. Stephans
ISBN 0-471-44454-5   Copyright © 2004 John Wiley & Sons, Inc.

Input Requirements

Input requirements for the system safety program plan are available project description data and requirements, provisions, or guidance provided externally and/or from internal system safety planning procedures.

Report Format

System Safety Organization. The names, titles, organizations, addresses, and telephone numbers of persons assigned to the SSWG, individuals in the review cycle, key support personnel, and other individuals involved in the system safety effort are listed.

System Safety Milestones. A list of major tasks with accompanying suspense dates, it includes life cycle phase start and stop dates, dates when specified system safety products are due, dates of formal reviews, and any other relevant dates.

System Safety Requirements. Specify the system safety products to be produced, the risk assessment code matrix to be used, risk acceptability criteria, and residual risk acceptance procedures.

Hazard Analyses. Specify the analytical tool and techniques to be employed.

System Safety Data. Specify the types of data that are required to support the effort, including sources of the data and information needed to acquire or access the data. Outline specific data generation and reporting requirements.

Safety Verification. Outline the safety verification and tracking procedures to be followed.

Training. Specify the training requirements for individuals involved in the system safety effort.

Other. Provide any other information to aid the system safety working group in accomplishing primary tasks. Information could be included on how the program will be monitored and audited, accident or problem reporting procedures, and support resources available, to include procedures for interfacing with support personnel or other system safety working groups.

Comments

The SSPP format outline parallels the format listed in data item description DI-SAFT-80100 and thus generally conforms to the requirements of MIL-STD-882B.

PRELIMINARY HAZARD LIST

Purpose

The preliminary hazard list (PHL) documents and provides initial assessment of hazards identified very early in the life cycle. The PHL is a feeder document for the preliminary hazard analysis (PHA) and provides the first information to aid in scaling the system safety effort.

Time of Preparation

The PHL should be started early in the concept phase and completed and delivered by the end of the concept phase.

Preparer

The PHL is normally prepared by the system safety working group and reviewed by management review boards.

Input Requirements

All available project description documents and relevant historical data are used in preparing the PHL.

Techniques

Informal conferencing, checklists, and energy trace and barrier analysis (ETBA) are techniques recommended for preparation of PHLs.

Report Format

The PHL report is generally a narrative with a summary that highlights significant risks and recommendations. It should also contain a brief project description, a discussion of the PHL methodology used including the risk assessment matrix, requirements for additional information, recommendations for follow-on studies or analyses, and the PHL worksheets.

Worksheet Format

Figure 8-1 shows a PHL worksheet.

Instructions

"Hazardous Event" column: A description of the hazards and/or undesired or unacceptable occurrences.

PRELIMINARY HAZARD LIST

Date _____
Page ___ of ___

Project _____

Prepared by _____

Method(s) used: ☐ Informal Conferencing   ☐ Checklist Review   ☐ ETBA ☐ Other _____

HAZARDOUS EVENT	CAUSAL FACTORS	SYSTEM EFFECTS	RAC	COMMENTS

Figure 8-1   Preliminary Hazard List (PHL).

"Causal Factors" column: A description of why or how the hazard may result in a mishap.

"System Effects" column: A description of each significant hazard that addresses how many people could be affected, how much is known about the hazard, how the community or the environment could be affected, and how the total system or other subsystems could be affected.

"RAC" column: The risk assessment code assigned to each uncontrolled hazard or undesired or unacceptable occurrence.

"Comments" column: Provision for comments by reviewers such as control recommendations or information on applicable codes, standards, or regulations.

PRELIMINARY HAZARD ANALYSIS

Purpose

The preliminary hazard analysis (PHA) expands the preliminary hazard list by identifying additional hazards, analyzing identified hazards, recommending hazard controls, and determining the level of risk after the controls are applied. The PHA results aid in determining the level of effort required for other hazard analyses.

Time of Preparation

The PHA is prepared as early as practical during the design phase. It should be developed in conjunction with the initial design and with close coordination between the system safety working group and the design group. Normally the PHA should be completed and delivered by the time the design is approximately 35% complete.

Preparer

The PHA is normally prepared by the system safety working group and reviewed by management review boards.

Input Requirements

Available project description information, historical data, and the PHL are used in preparing the PHA.

Techniques

Failure modes and effects analysis (FMEA) and energy trace and barrier analysis are the techniques recommended to aid in conducting PHAs.

Report Format

The PHL report is generally a narrative with a summary that highlights significant risks and recommendations. It should also contain a brief project description, a discussion of the PHL methodology used, including the risk assessment matrix, requirements for additional information, recommendations for follow-on studies or analyses, and the PHA worksheets.

Worksheet Format

A preliminary hazard analysis worksheet is shown in Figure 8-2.

Instructions

"Recommended Actions" column: Describe the actions recommended to control the hazard.

"Controlled RAC" column: Enter the risk assessment code after the recommended controls are in place.

"Standards" column: Enter any relevant codes, standards, or regulations.

PRELIMINARY HAZARD ANALYSIS

Project _____

Prepared by _____

Method(s) used: ☐ ETBA ☐ FMEA ☐ FTA ☐ Other _____

Date _____
Page ___ of ___

HAZARDOUS EVENT	CAUSAL FACTORS	SYSTEM EFFECTS	RWC	COMMENTS	RECOMMENDED ACTIONS	CONTROLLED RMC	STANDARDS

Figure 8-2 Preliminary Hazard Analysis (PHA).

HAZARD TRACKING LOG

Purpose

The hazard tracking log provides a means of tracking hazards throughout the life cycle and documenting the status of corrective actions.

Time of Preparation

The hazard tracking log is prepared as an add-on to the PHA and prepared as soon as the PHA is completed and delivered. The hazard tracking log is then updated as control actions are taken and verified. Summary reports of hazard tracking status are prepared and available for all formal system safety reviews.

Preparer

The hazard tracking log is normally prepared and maintained by the safety staff.

Input Requirements

Because the hazard tracking log is an extension of the preliminary hazard analysis, the PHA must be complete before the hazard tracking log is needed. Information needed to complete the hazard tracking log includes appropriate drawing numbers, change orders, and other documents that verify that hazard controls have been implemented.

Techniques

Hazard tracking documentation is normally accomplished by close coordination between the safety staff and those responsible for implementing hazard controls, usually the designers during the design phase, production or construction supervision during the fabrication phase, and operations or maintenance supervision during the operations phase. A suspense file and formal reporting form are recommended to provide backup documentation.

Report Format

A hazard tracking summary is prepared and available for all formal system safety review meetings. The summary should list all open (uncontrolled) hazards in chronological order and/or by RAC. This type of report should also be available on an as-needed basis.

Worksheet Format

Figure 8-3 shows a hazard tracking log incorporated in a preliminary hazard analysis worksheet.

Instructions

"Action Taken" column: Describes the nature of the action taken to eliminate or control the hazard.

"Reference Document" column: Lists the drawing and/or other hazard tracking reference that documents the corrective action(s).

"Verification Signature(s)" column: Contains the signature of the person(s) responsible for maintaining the hazard tracking log and the date the verification was made.

SUBSYSTEM HAZARD ANALYSIS

Purpose

The subsystem hazard analysis (SSHA) provides detailed analysis of hazards associated with specific systems. Depending upon the complexity and nature

PRELIMINARY HAZARD ANALYSIS
WITH HAZARD TRACKING LOG

Date _____
Page ___ of ___

Project _____

Prepared by _____

Method(s) used: ☐ETBA ☐ FMEA ☐ FTA ☐ Other _____

HAZARDOUS EVENT	CAUSAL FACTORS	SYSTEM EFFECTS	RAC	COMMENTS	RECOMMENDED ACTIONS	CONTROL-LED RAC	STANDARDS	ACTION TAKEN	REFERENCE DOCUMENT	VERIFICATION SIGNATURE(S)

Figure 8-3   Hazard Tracking Log Worksheet.

of the end product, SSHAs may be conducted for all subsystems, only specified subsystems, or no subsystems.

Time of Preparation

The SSHA is prepared as early in the design phase as possible, generally
between 35% and 60% design. Timing will normally be determined by the
availability of subsystem design data.

Preparer

The SSHA is normally prepared by the system safety working group and
reviewed by management review boards.

Input Requirements

Detailed project description documents and drawings are required to perform
an SSHA. Additionally, relevant codes, standards, and regulations, the PHL,
and the PHA for the projects should be available. Access to "lessons learned"
and reliability data is also needed.

Techniques

Failure modes and effects analysis (FMEA) and fault tree analysis (FTA) are the techniques recommended for SSHAs. Energy trace and barrier analysis (ETBA) may also be appropriate. Change analysis is useful for evaluating changes made during the SSHA.

Report Format

The subsystem hazard analysis report contains a description of the subsystem and a narrative summary of key findings that specifically address the adequacy of the controls placed on any high hazards associated with the end products, the level of residual risks that remain after controls have been applied, and recommendations for further analysis or testing. The report should also describe the techniques and methodology used in performing the analysis, including risk assessment and risk acceptance criteria. The report should also contain the hazard report worksheets used in the study.

Worksheet Format

Figure 8-4 shows a subsystem hazard analysis worksheet.

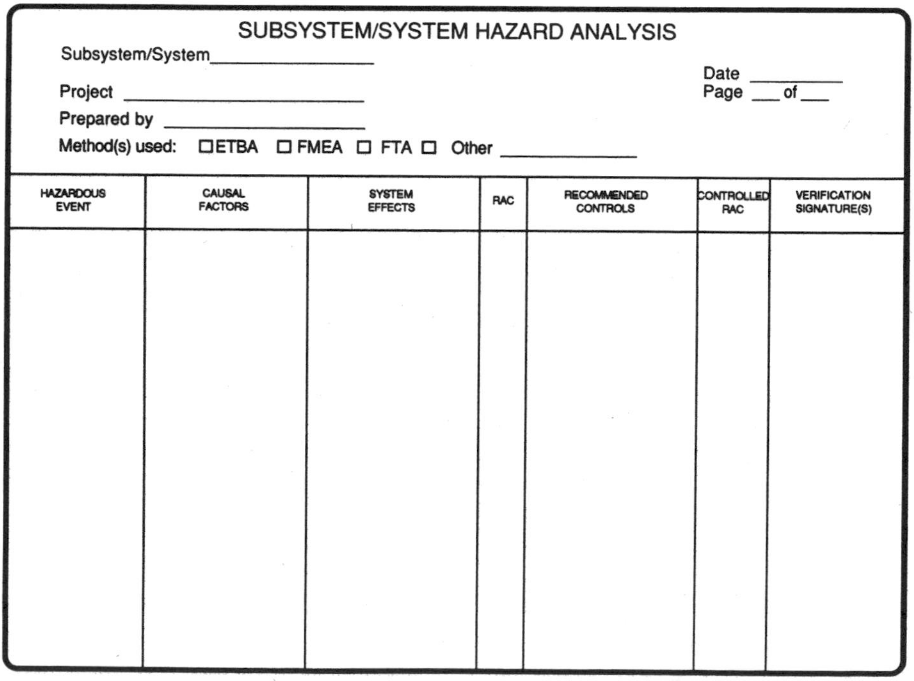

SUBSYSTEM/SYSTEM HAZARD ANALYSIS

Subsystem/System _____

Project _____   Date _____

Prepared by _____   Page ___ of ___

Method(s) used: ☐ETBA ☐ FMEA ☐ FTA ☐ Other _____

HAZARDOUS EVENT	CAUSAL FACTORS	SYSTEM EFFECTS	RAC	RECOMMENDED CONTROLS	CONTROLLED RAC	VERIFICATION SIGNATURE(S)

Figure 8-4   Subsystem/System Hazard Analysis Worksheet.

Instructions

"Hazardous Event" column: A description of the hazard (usually associated with an energy source or flow of energy).

"Causal Factors" column: A description of why or how the hazard may result in an accident.

"System Effects" column: A description of the potential accident that could result from the hazard in terms of number of persons affected, environmental impact, and effects on other subsystems or the total system.

"RAC' column: The risk assessment code assigned to the hazard.

"Recommended Controls" column: Recommendations for eliminating the hazard or controlling it by reducing the probability and/or severity of an accident resulting from the hazard.

"Controlled RAC" column: The risk assessment code assigned to the hazard after the recommended controls are in place.

"Verification Signature(s)" column: The signature of the responsible person(s) verifying that the control is in place and is adequate and/or the signature of the person(s) accepting the residual risk. (Signature blocks include printed or typed names and dates.)

SYSTEM HAZARD ANALYSIS

Purpose

The system hazard analysis (SHA) provides detailed analysis of hazards associated with the total system.

Time of Preparation

The SHA is prepared as early in the design phase as possible, generally between 35% and 90% design. Timing is normally determined by the availability of subsystem and system design data.

Preparer

The SHA is normally prepared by the system safety working group and reviewed by management review boards.

Input Requirements

Detailed project description documents and drawings are required to perform an SHA. Additionally, relevant codes, standards, and regulations, the PHL, the PHA, and the SSHAs for the projects should be available. Access to lessons learned and reliability data is also needed.

Techniques

Failure modes and effects analysis (FMEA) and fault tree analysis (FTA) are recommended techniques for the SHA. Energy trace and barrier analysis (ETBA) may also be appropriate. Change analysis is useful for evaluating changes made during the SHA.

Report Format

The SHA report contains a description of the system and a narrative summary of key findings that specifically address the adequacy of the controls placed on any high hazards associated with the end products, the level of residual risks that remain after controls have been applied, and recommendations for further analysis or testing. The report also describes the techniques and methodology used in performing the analysis, including risk assessment and risk acceptance criteria. The report also contains the hazard report worksheets used in the study.

Worksheet Format

Figure 8-4 shows a system hazard analysis worksheet.

Instructions

"Hazardous Events" column: A description of the hazard (usually associated with an energy source or flow of energy).

"Causal Factors" column: A description of why or how the hazard may result in an accident.

"System Effects" column: A description of the potential accident that could result from the hazard in terms of number of persons affected, environmental impact, and how other subsystems or the total system could be affected.

"RAC" column: The risk assessment code assigned to the hazard.

"Recommended Controls" column: Recommendations for eliminating the hazard or controlling it by reducing the probability and/or severity of an accident resulting from the hazard.

"Controlled RAC" column: The risk assessment code assigned to the hazard after the recommended controls are in place.

"Verification Signature(s)" column: The signature of the responsible person verifying that the control is in place and is adequate and/or the signature of the person accepting the residual risk. (Signature blocks include printed or typed names and dates.)

OPERATING HAZARD ANALYSIS

Purpose

The operating hazard analysis (OHA) analyzes hazards associated with the maintenance and operation of the system, with emphasis on human factors, training and procedures, and the person-machine interface. The OHA is sometimes called the O&SHA (operating and support hazard analysis).

Time of Preparation

The initial OHA should be prepared as early in the life cycle as possible. Because the design must be nearly complete and operating and maintenance procedures must be available (at least in draft), however, the OHA is usually started late in the design phase or early in the production phase. It is completed before the end of the production phase. The timing of the OHA is highly dependent on the nature of the end product. If the end product is simply an updated version of an existing system, many applicable operating and maintenance procedures may be available early in the project, and the nature of the end user organization may already be defined. In this situation, the pacing item is the availability of system design information, but the OHA can be initiated relatively early in the design phase and completed or nearly completed by the end of the design phase. For example, if the end item is a facility that is to replace an existing facility with the existing work force and equipment performing the same tasks in the new facility, the OHA can be performed during design. For a totally new system, especially one that involves production and testing of prototypes, however, the OHA may be initiated late in the design phase and prepared primarily during the production phase. Update OHAs are performed periodically throughout the life cycle.

Preparer

The OHA is normally prepared by the system safety working group and reviewed by management review boards.

Input Requirements

Detailed project description and design information; applicable codes, standards, and regulations; the PHL, PHA, SSHAs and SHA; and operating and maintenance procedures (draft or final) are needed to perform an OHA. Detailed equipment envelopes and data about the size, composition, skill levels, and organization of the personnel operating and maintaining the system are also required.

Techniques

Project evaluation tree (PET) analysis is recommended. Procedural audits, ETBA, and change analysis may also be appropriate.

Report Format

The OHA report contains a description of the system including operating and maintenance organizations and procedures, as well as a narrative summary of key findings and recommended controls. A schedule for recommended update OHAs is also included.

The OHA worksheet format and instructions for completing it are included in the explanation of PET analysis in Chapter 16.

CHANGE ANALYSIS REPORT

Purpose

The change analysis report provides formal documentation and feedback of safety analyses performed on changes to the end product throughout the life cycle.

Time of Preparation

A change analysis report is prepared at any time during the life cycle when a change is made to a design, drawing, component, subsystem, or any other part of the project after the safety analysis of that particular part of the project is complete.

Preparer

The change analysis report is normally prepared by the system safety working group.

Input Requirements

Detailed descriptions of the changes and of the parts of the end product affected by the changes are required.

Techniques

Change analysis is the recommended technique.

Report Format

A narrative containing a summary of the changes, the possible problems or hazards created by the changes, and recommended counter-changes or controls is normally all that is required for relatively simple changes. Complex changes that create significant hazards may require the use of hazard analysis worksheets. The change analysis report should also recommend further analyses if appropriate.

Worksheet Format

Worksheets are normally not required. If the complexity of the change dictates the use of worksheets, use the system/subsystem hazard analysis worksheet format depicted earlier in Figure 8-4.

Instructions for preparing a change analysis report are included in Chapter 17.

ACCIDENT ANALYSIS REPORT

Purpose

The accident analysis report determines and documents the root causes of accidents associated with the end product and includes new hazards, hazards inadequately controlled or analyzed, and new baseline information identified by the accident analysis in the system safety effort.

Time of Preparation

Accident analysis reports are prepared in response to all significant accidents, including near misses, associated with the end product at any time during the life cycle.

Preparer

An accident analysis report may be prepared by the system safety working group and/or an independent accident investigation board. The recommended method for serious accidents is for the SSWG to prepare an accident analysis report based 'on its own investigation and upon the findings of the traditional accident report prepared by the accident investigation board.

Input Requirements

Detailed project description and detailed accident information are required to conduct an in-depth accident analysis. Historical data, existing hazard analyses, and the hazard tracking log are also important.

Techniques

Techniques recommended for conducting accident analysis include change analysis, event and causal factors charts, PET analysis (or MORT or mini-

MORT analysis), extreme value projection, and, if applicable, time-loss analysis.

Report Format

The accident analysis report is a narrative. It includes a description of the project and the accident, a summary of findings and recommendations, an outline of methodology, and a detailed discussion of facts, findings, and recommendations. Photographs, drawings, and an event and causal factors chart are included as illustrations.

Worksheet Format

Generally worksheets are not required, even though hazard analysis worksheets may be prepared to identify and analyze new hazards identified in the analysis and/or hazards not adequately studied during earlier analysis efforts.

Instructions for the techniques recommended for accident analysis are included in Part 4.

REVIEW QUESTIONS

1. Who generally prepares the system safety program plan (SSPP)?
2. Who prepares the preliminary hazard list? When? What techniques are recommended?
3. What are the input requirements for preliminary hazard analysis (PHA)? What techniques are recommended?
4. What columns are added to the PHA to form the hazard tracking log?
5. When should subsystem and system hazard analyses be performed? By whom? What are the input requirements?
6. When should a change analysis report be prepared?
7. What techniques are recommended for conducting an accident analysis?
8. What techniques are recommended for conducting an operating hazard analysis? When is it performed? By whom?

CHAPTER 9

Program Implementation

To initiate a system safety program, a number of planning steps must be completed. This process in influenced by the nature of the organization and the end product, but the following steps provide a general, generic approach that, with some editing, can apply to most situations.

Step 1: Outline purpose, objectives, milestones, and resource requirements

Develop a clear reason for having a system safety effort within the organization and objectives for the program.

If you are establishing the program in response to government (or other) contract requirements, the stated purpose of the program can be: "To produce the safest end product possible within operational, budget, and other constraints" (MIL-STD-882B terminology). An obvious system safety program objective would be: "To fulfill contract requirements."

If you are establishing a program internally, solely as a means of improving your overall safety effort, the overall purpose migh be: "To systematically apply state-of-the-art hazard identification, analysis, and control tools and techniques as early in the life cycle as practical to cost-effectively provide significant improvements in the safety effort," "To cost-effectively provide significant safety and design improvements," or "To provide a better product at a lower cost."

Program objectives could include:

- To reduce product recalls or retrofits
- To reduce safety-related losses
- To ensure that safety is incorporated during design

Establish milestones for completing the steps necessary to establish the program and provide an initial estimate of resources required to plan and to implement the program. At this point, all planning must be very general in nature. The purpose, objectives, milestones, and resource requirements will be

System Safety for the 21ˢᵗ Century: The Updated and Revised Edition of System Safety 2000, by Richard A. Stephans
ISBN 0-471-44454-5 Copyright © 2004 John Wiley & Sons, Inc.

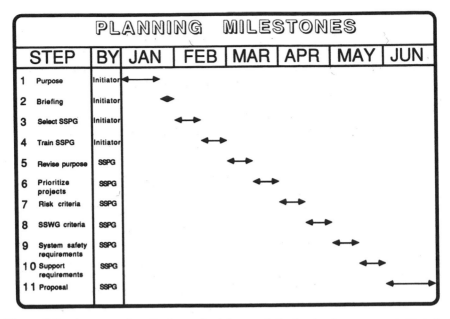

Figure 9-1   An example milestone chart for a relatively simple program initiated at the first of the calendar year.

developed in more detail by the system safety planning group (SSPG) to be selected in Step 3.

In most cases, a system safety effort can be planned and implemented within six months. A recommended milestone chart is shown in Figure 9-1.

Resource requirements for planning and implementation generally consist of person-hours of effort by the SSPG. Assuming the SSPG consists of four people who meet two hours every two weeks for six months, the total SSPG effort would be about 100 hours. Add a two-day training session for the SSPG and the support services provided by professional and clerical members of the safety staff, and the total planning effort would require approximately 560 management and engineering hours and 100 clerical hours. In the long run, the program should produce significant cost savings.

Step 2: Brief management

At this point in most organizations, conducting a management briefing to obtain approval to form a system safety planning group and proceed with the project would be appropriate. The nature of the briefing or approval process, of course, depends on the position and authority of the individual attempting to initiate the program and his or her relationship with higher management. In some situations, this step may not be necessary at all; in others, it may require obtaining support from multiple levels and present a significant task.

Step 3: Select planning group

A successful system safety program depends on support and resources from top management, project management, engineering, safety, operations, maintenance, and perhaps other areas like legal, procurement, and marketing.

Sooner or later, selling them and perhaps other people in the organization on the benefits and value of the program becomes necessary.

Frequently, early involvement by key personnel produces a program that is easy to sell because (1) those involved in developing the program identify with it and tend to support it, and (2) more early input by more people produces a good program.

As a matter of fact, the whole concept of system safety is that early input by more people ultimately results in a better product. Selling that concept may be difficult if your system safety plan is developed in a vacuum.

In selecting persons for this group, attempt to select individuals who can make meaningful contributions. Basically, there are innovators and there are imitators. At this point, you need to work with innovators.

Forming a planning group involves many variables. Large groups tend to work less quickly and less cost-effectively but provide a larger base of support and ultimately produce better products. Small groups (or individuals) tend to work more rapidly and efficiently but may not consider as many factors and thus produce programs that are less acceptable to those who were left out of the planning process. At a minimum, representatives from the *safety, engineering, management,* and *end user* communities should be given the opportunity to provide input during the planning process; note that these are the same core groups that should be included in system safety working groups.

System safety efforts that are mandated by contract can, of course, be imposed on the organization. However, the quality of programs imposed on an organization with a single purpose of meeting contract requirements is questionable. The quality of a program that is tolerated and supported only to the extent of meeting minimum contract requirements may be much different from a program the organization perceives as a worthwhile, cost-effective means of producing a better product.

Even system safety efforts required by contract should include appropriate elements of the organization in early planning stages.

Members of the system safety planning group normally become members of a system safety working group or safety review board.

Step 4: Organize and train the system safety planning group

As with any group or committee, organizing to accomplish tasks can be done in a number of ways. All work can be done as a group. This method is usually the slowest and most tedious, but it ensures full participation and keeps all group members informed on all tasks.

Another approach is to parcel out tasks to individuals within the group and then review progress on each task as a group. This approach is generally more efficient but may limit valuable interaction. Generally, a combination of the methods is recommended for performing system safety planning tasks, with group versus individual tasks selected very carefully.

To function effectively in planning a system safety program, all members of the group should understand the fundamentals of system safety. This book can provide the basis for developing the training programs required to support the program. For example, a general SSPG training outline would resemble the following:

Day One
Morning—Introduction to systems safety (overview of Part 1)
Afternoon—System safety program planning and management (overview of Part 2)
Day Two
Morning—Analytical aids (overview of Part 3)
Afternoon—System safety analysis techniques (overview of Part 4).

The initial program established to provide an overview of system safety to planning group members should evolve into an overview of system safety for managers, engineers, and supervisors. One of the tasks of the planning group is to determine other system safety training requirements.

The following is a sample system safety planning group (SSPG) organization with recommended responsibilities. (Note: Circumstances may make it desirable to tailor the composition of the SSPG and/or individual responsibilities.)

Management Representative: Serves as head of the planning group, conducts meetings, and assigns tasks. Serves as the primary spokesperson for the group to top management. Provides project objectives and constraints information. Primary responsibility for providing system safety program resources.

Safety Representative: Serves as executive secretary of the SSPG. Keeps minutes of meetings and documents group tasks. Primary responsibility for training and administrative matters to include (with concurrence of the management representative) scheduling SSPG meetings and setting agendas. Represents the initiator of the program and the staff function with primary responsibility for overseeing the system safety effort. Primary responsibility for providing input relative to safety codes, standards, and regulations, safety historical data, and hazard analysis techniques. Provides interface with the safety community and, in the absence of the management representative, heads the SSPG.

Engineering Representative: Primary responsibility for providing project description and design information. Provides interface with the engineering community.

End User: In an in-house project, the end user representative is normally from the operations and/or maintenance organization. If the end product is a consumer good, the end user representative could be an actual consumer, a representative from a consumers' group, and/or organizational public relations or marketing personnel. In any event, the representative from the end user community should provide input relevant to operations and maintenance of the end product.

At this point, the SSPG becomes responsible for continued planning and completes steps 5 through 11.

Step 5: Refine program purpose, objectives, milestones, and resource requirements

One of the major tasks for the SSPG is to refine and update the program purpose, objectives, milestones, and resource requirements. Even though this task should continue throughout the planning process and may be one of the last planning tasks completed, it should be identified, addressed, and initiated early in the planning process (Fig. 9-2).

Step 6: Identify and prioritize projects for system safety efforts

An organization may benefit from two types of system safety programs. The first focuses on the end product designed, manufactured, and/or produced by

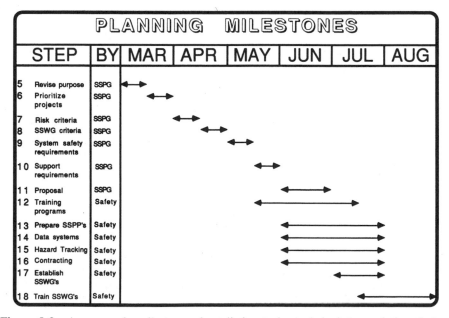

Figure 9-2   An example milestone chart listing tasks and depicting relative timing of planning steps performed by the SSPG and the support steps taken by the safety organization.

the organization. If the organization produces or designs only one end product, of course, no prioritization of the system safety effort is required. If, however, the organization designs and/or produces a variety of different end products, prioritizing products for early system safety efforts is probably prudent.

The second type of system safety program focuses on the organization's own facilities, equipment, or operations, not on its end products. A weapons system contractor, for example, may benefit from conducting a system safety program for a new production facility or even for selected items of plant equipment.

A system safety effort that has multiple targets—either end items, organizational facilities or equipment, or both—needs criteria for prioritizing system safety projects.

The most logical approach is to prioritize according to the total risk associated with each item. The total risk is determined by examining the probability and severity of loss for a given unit and multiplying by the total exposure in terms of units produced and project life. Methods of determining total risk are discussed in Chapter 11.

Step 7: Select risk assessment and risk acceptance methods and criteria
Review the Total Risk Exposure Code matrix and risk acceptance criteria in Chapter 11. Adopt the risk assessment and acceptance criteria and methodology (as presented or as modified by the planning group) or develop and adopt alternatives.

Ideally, one set of risk assessment and acceptance criteria and methods can be accepted and applied to all projects. However, contractual requirements or other constraints or considerations may require that different criteria and/or method be applied to different projects.

Step 8: Select system safety working group criteria
Normally, a separate system safety working group (SSWG) is designated for each project. In some situations, however, one SSWG may successfully participate in multiple projects.

The suggested composition of the SSWG is as follows:

Project Manager: The head of the SSWG is normally the project manager, the individual with overall safety responsibility for the project. The project manager provides specific details about the project, including objectives and constraints, and brings organizing, planning, directing, leading, and controlling skills to the effort. The project manager should be the primary interface with higher levels of management.

Safety Engineer or Professional: The safety representative serves as the executive secretary for the SSWG. He or she provides or coordinates administrative support for the board and brings a knowledge of safety codes, standards, and regulations and hazard identification, analysis, and control techniques. The safety representative has primary responsibility for providing or coordinating system safety support services, including establishing and maintaining a comprehensive system safety training program, a system safety data collection and retrieval system, and system safety records, including hazard tracking logs, as well as technical oversight of system safety contracting efforts. The safety representative is also the primary interface with other members of the safety community.

Project or Design Engineer: Those who have primary design responsibilities are included in the SSWG. The design engineer representative provides detailed project descriptions and has the engineering skills necessary to aid in eliminating or controlling hazards associated with the project.

End User Representative: As with the planning group representative, representation from this group is dependent upon the nature of the end products. For an in-house project, the end user representative(s) would normally be from the operations and/or maintenance organization. If the end product is a consumer good, the end user representative could be a consumer, a representative from a consumers' group, and/or organizational public relations or marketing personnel. In any event, the representative from the end user community should provide input relevant to operations and maintenance of the end product. The special knowledge and input from the end user representative includes a knowledge of the real requirements for the end product from a user's point of view. This SSWG member provides key human factors information needed to evaluate procedures, instructions, training requirements, and other human interface concerns. In some instances, the end user representative is not a member of the organization conducting the system safety program and has to be "borrowed" from another organization and/or hired as a consultant. However it is obtained, input from the end user is critical to the system safety effort.

Step 9: Establish system safety requirements
The system safety planning group either adopts the tasks and products outlined in Chapters 7 and 8 or develops alternatives. This effort should include guidelines for establishing project phases, review points, and levels of review and approval.

Step 10: Establish system safety support requirements
The system safety planning group should provide guidance in developing and maintaining a comprehensive system safety training program, a data collection and retrieval system, and a hazard tracking program. The safety organization normally has the responsibility for providing these support services. Additionally, the safety organization serves as the technical evaluator and technical point of contact for system safety contractors' activities.

Step 11: Seek management approval of system safety program guidelines

The SSPG prepares a written report outlining the system safety program guidelines, briefs management on the guidelines, and obtains approval to turn the guidelines over to the safety organization to develop individual system safety program plans and administer the system safety effort. This step completes the initial planning tasks of the system safety planning group. This group would normally become the first level of review for system safety program plans prepared by the safety organization and for system safety products produced by the SSWGs.

Step 12: Establish a comprehensive system safety training program

The safety organization establishes a comprehensive system safety training program. At a minimum, the following courses are recommended:

Management Overview: A two-hour class for corporate management, department managers, and key staff members. The first hour of the course would be an introduction to system safety, including a brief history, basic concepts, organization, tasks, and products. The second hour would focus on system safety program planning and management.

System Safety Working Group Workshop: A 4.5-day program for members of SSWGs. (See appendix for course outline and Figure 9-3 for course schedule.)

	MONDAY	TUESDAY	WEDNESDAY	THURSDAY	FRIDAY
1ST HOUR	COURSE INTRODUCTION	PE PRESENTATIONS	PE PRESENTATIONS	PE PRESENTATIONS	PE PRESENTATIONS
2ND HOUR	INTRODUCTION TO SYSTEM SAFETY (CHAPTERS 1-7)	PRELIMINARY HAZARD LIST (PHL) (CHAPTER 8)	HAZARD TRACKING & VERIFICATION (CHAPTER 8)	OPERATING HAZARD ANALYSIS (CHAPTER 8)	PROGRAM IMPLEMENTATION (CHAPTER 9)
3RD HOUR		PRELIMINARY HAZARD ANALYSIS (CHAPTER 8)	SUBSYSTEM & SYSTEM HAZARD ANALYSIS (CHPT 8)	CHANGE ANALYSIS (CHAPTERS 8,17)	
4TH HOUR		HUMAN FACTORS (CHAPTER 12)	FAILURE MODES & EFFECTS ANALYSIS (CHAPTER 14)	ACCIDENT ANALYSIS (CHAPTERS 8,17)	COURSE CRITIQUE/ CLOSING
	LUNCH				
5TH HOUR	RISK ASSESSMENT CODES (CHAPTER 11)	ENERGY TRACE & BARRIER ANALYSIS (CHAPTER 13)	FAULT TREE ANALYSIS (CHAPTER 15)	PET ANALYSIS (CHAPTER 16)	
6TH HOUR	ANALYTICAL TREES (CHAPTER 10)				
7TH HOUR	PE ANALYTICAL TREES	PE PHL/PHA	PE SSHA/SHA	PE OHA	
8TH HOUR					

Figure 9-3 System safety working group workshop. Course schedule for workshop presented by safety organization to train SSWG members.

	MONDAY	TUESDAY	WEDNESDAY	THURSDAY	FRIDAY
1ST HOUR	COURSE INTRODUCTION	PE PRESENTATIONS	PE PRESENTATIONS	PE PRESENTATIONS	PE PRESENTATIONS
2ND HOUR	INTRODUCTION TO HAZARD ANALYSIS	FAILURE MODES & EFFECTS ANALYSIS (CHAPTER 14)	PET ANALYSIS (CHAPTER 16)	ACCIDENT ANALYSIS (CHAPTERS 8,17)	HAZARD REPORT WRITING
3RD HOUR	(CHAPTERS 13-17)	PE FMEA	PE PET	PE ACCIDENT ANALYSIS	COURSE REVIEW
4TH HOUR	RISK ASSESSMENT CODES (CHAPTER 11)				COURSE CRITIQUE/ CLOSING
	LUNCH				
5TH HOUR	ENERGY TRACE & BARRIER ANALYSIS (CHAPTER 13)	PE PRESENTATIONS	PE PRESENTATIONS	PE PRESENTATIONS	
6TH HOUR		FAULT TREE ANALYSIS (CHAPTER 15)	CHANGE ANALYSIS (CHAPTER 17)	OTHER ANALYSES (CHAPTER 17)	
7TH HOUR	PE ETBA	PE FTA	PE CHANGE ANALYSIS	PE OTHER ANALYSES	
8TH HOUR					

Figure 9-4 Hazard analysis techniques workshop. Course schedule for workshop recommended for safety professionals and engineers.

Hazard Analysis Techniques for Engineers: A 4.5-day workshop for project, safety, and design engineers (See appendix for course outline and Figure 9-4 for course schedule.)

Step 13: Prepare system safety program plans
Using the guidance provide by the SSPG, the safety organization should prepare a system safety program plan (SSPP) for each project. The SSPP should be prepared using the format outlined in Chapter 8 or as prescribed by the SSPG. The SSPPs should be reviewed and approved by the SSPG or as specified by SSPG guidance.

Step 14: Develop data requirements
Based upon the guidance provided by the SSPG, the safety organization should establish a comprehensive data collection and retrieval system.

Ideally, internal accident, incident, and close call information should be augmented by appropriate national data banks such as the Government Industry Data Exchange Program (GIDEP), the Remote Console (RECON) system that accesses the Department of Energy Technical Information Center, the U.S. Army Corps of Engineers Facility System Safety Information Center (FASSIC), and/or any of the commercial data banks providing safety or reliability data.

Data should be easy to access and sort. Sort fields could include type of part or product, manufacturer, cost, size, use, types of accidents or failures, geographic location, time, date, temperature, other environmental conditions, types of injuries, energy sources, dollar loss, analyst, project name, project number, system, subsystem, risk assessment code, and any other relevant comments that may aid in analysis of data.

Menu-driven, keyword programs, or other user-friendly systems are recommended.

Step 15: Establish hazard tracking and documentation
The safety organization supports the system safety working group (SSWG) by maintaining a current hazard tracking log for each project. Ideally, the system can be preprogrammed with review schedules and milestones and automatically generate reports to aid SSWGs in tracking and verifying corrective actions. The system should be capable at generating reports on demand, as well as reports specifically generated to document a specific review point.

Step 16: Establish contracting procedures
If the end product (system) is not an in-house effort, aid procurement personnel in writing system safety requirements into requests for proposals and statements of work and in evaluating the system safety portions of technical proposals.

Aid contractors and contract administrators in understanding system safety requirements.

If a consultant is required to aid in the system safety effort, aid in preparing the scope of work for the effort. Consultants can be helpful in establishing or validating new programs by performing analyses, serving as a member of a system safety working group or system safety planning group, serving as an advisor or review agent, or conducting system safety training.

Step 17: Establish system safety working groups
The safety organization, in coordination with the groups providing its members, designates system safety working groups for the projects recommended by the system safety planning group. The number and size of the groups depend upon the size of the organization and the number of projects (end products) involved. The size of the group should generally range from four to twelve persons. Complex systems may have separate SSWGs for major subsystems.

Step 18: Train system safety working groups
Members (and potential members) of system safety working groups attend the SSWG workshop (see Step 12) before implementing system safety program plans. Additionally, the safety and engineering representatives in each SSWG complete the hazard analysis workshop (see step 12). Training is scheduled and conducted by the safety organization.

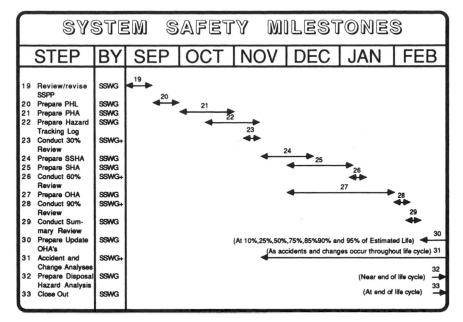

Figure 9-5   An example of a continuation milestone chart depicting system safety tasks and relative timing for a simple system safety program. *SSWG+* indicates that persons other than, or in addition to, the SSWG may be involved.

Step 19: Review and/or revise phases, milestones, and review points
The SSWG reviews the system safety program plan and verifies that the phases, milestones, and review points are accurate and adequate (Fig. 9-5). If necessary, these items are revised.

Step 20: Prepare PL
The SSWG prepares a preliminary hazard list for the project as described in Chapter 8 (or prescribed in the system safety program plan).

Step 21: Prepare PHA
The system safety working group expands the PHL into a PHA as described in Chapter 8 (or as outlined in the system safety program plan).

Step 22: Prepare hazard tracking log
Based on the PHA, the safety organization prepares a hazard tracking log as outlined in Chapter 8 (or as described in the system safety program plan).

Step 23: Conduct 30% design review
The PHL, PHA, and hazard tracking log are reviewed by the appropriate review boards relatively early in the design phase (as prescribed in the SSPP). At a minimum, the manager or supervisor of each member of the SSWG (for

example, the safety manager and the engineering manager) should be in the review cycle.

Step 24: Prepare subsystem hazard analysis
Based on the results of the PHA, recommendations made by 30% review boards, and guidance provided in the system safety program plan, detailed hazard analyses are made of specified (critical) subsystems. The techniques for these SSHAs are as outlined in the system safety program plan or as selected by the SSWG. Failure modes and effects analysis (FMEA) and/or fault tree analysis (FTA) are generally the techniques of choice. Software hazard analysis, common cause analysis, and/or sneak circuit analysis may also be appropriate.

Step 25: Prepare system hazard analysis
Based on the system safety program plan and recommendations from other analyses (PHA, SSHAs), a system hazard analysis (SHA) may be required. The SHA is initiated as soon as input data from the SSHAs are available and completed at the review milestone indicated in the SSPP, normally around 60% design.

Step 26: Conduct 60% design review
The PHL, PHA, SSHAs, SHA, and hazard tracking log are reviewed by the appropriate review boards relatively early in the design phase (as prescribed in the system safety program plan). At a minimum, the manager or supervisor of each member of the SSWG (for example, general manager, safety manager) is in the review cycle.

Step 27: Prepare initial OHA
As soon as practical, consistent with SSPP guidance, an OHA is prepared by the SSWG. Project evaluation tree (PET) analysis is the recommended technique.

Step 28: Conduct 90% design review
The PHL, PHA, SSHAs, SHA, OHA, and hazard tracking log should be reviewed by the appropriate review boards relatively early in the design phase (as prescribed in the system safety program plan). At a minimum, the manager or supervisor of each member of the SSWG should be in the review cycle. (for example, the general manager, safety manager, and engineering manager).

Step 29: Conduct summary review of all design and production changes
Even though all changes should be reviewed and analyzed as they occur, the SSWG conducts a summary review of all changes made during the design and production stages to ensure that safety has not been compromised. This review is completed near the end of the project.

Step 30: Update OHAs (10%, 25%, 50%, 75%, 85%, 90%, and 95% of estimated life)

As previously noted, many current system safety efforts tend to address only the acquisition life cycle, that is, the phases of the life cycle dedicated to getting the end product completed or into full production.

The upstream system safety effort is clearly an extremely important part of an overall program, but significant contributions can and should also be made during the operating life of the end product.

A number of variables were considered during the design and testing of the end product. The system safety effort during the operating phase piodically compares actual parameters with those used in earlier analyses, updates operating and maintenance procedures as needed, and, if necessary, modifies the end product.

The most logical analysis to update is the operating hazard analysis. Updates can generally be accomplished by conducting a new OHA at each update point and/or by conducting a comprehensive change analysis to evaluate the differences in procedures, personnel, plant, and hardware between the original analysis (or last update) and the current analysis.

Not only changes in variables and associated risks are examined but also the baseline assumptions made in the risk assessment process (total exposure, number of products, product life, failure rates, and so on) are validated or adjusted, as well the risk acceptability criteria.

Operating hazard analysis update points ought to be established in the original OHA. The parameter for establishing the estimated life can be hours of operation, cycles, calendar years, or some combination of these or other factors. Once established in terms of a percentage of estimated life, the update points are not modified (without compelling reasons). The actual date for conducting each update analysis is established using the new "estimated life" as determined by the most recent update.

For example, a facility originally designed to have a useful life of forty years becomes operational in 1990. The 10% update (conducted in 1994, four years after the facility is opened) does not change the estimated life. However, the 25% update OHA (conducted in 2000) estimates that because of a combination of poor maintenance, changes in occupancy and use, and other factors, the building will serve for only another 20 years; that is, the estimated life is changed from 40 years to 30 years. The 50% update OHA, which would have been performed in 2010 based on original estimates, should now be performed in 2005.

The primary purpose of each update OHA is to identify, analyze, and control hazards associated with the operation of the end product. In order to do so properly, baseline data (risk assessment and acceptance criteria, estimated failure rates, exposure, and the like) must be reviewed and revised as indicated by the most current information available.

Update OHAs should be scheduled to address the "bathtub curve" for the end product. The bathtub curve predicts that accident and incident rates tend to be high during initial operations when the product is new, drop off rapidly as oversights are corrected, remain relatively flat (hopefully low) during most of the product's life, and then increase toward the end of the life cycle as problems associated with wear-out surface.

Analyses are performed at 10%, 25%, 50%, 75%, 85%, 90%, and 95% of estimated product life or as otherwise determined.

The update OHA performed after the end product has been in service for 10% of its estimated life (or as otherwise scheduled) concentrates on validating/revising baseline data used in earlier analyses and addressing hazards associated with start-up operations and new products.

The OHA performed at each point concentrates on cumulative changes that have occurred since the last OHA to ensure that appropriate counter-changes have been made and to validate or revise baseline data.

Step 31: Accident and change analysis
All of the other steps in the system safety process are proactive. This step is reactive. It is important that whenever there is a major accident involving the end product an accident analysis be performed and that the lessons learned be fed back into the system safety effort.

Root causes must be identified and appropriate corrective action implemented. The accident analysis may be performed by the SSWG and/or by other boards of investigation, but the necessity for systematic, in-depth accident analysis is an integral part of the system safety effort.

Similarly, major changes or modifications to the end product should be formally analyzed as part of the system safety effort. Recommendations and corrective action resulting from accident or change analyses should not be documented, reviewed, or tracked differently than corrective actions from other system safety analyses.

Step 32: Prepare disposal hazard analysis
The final hazard analysis to be performed by the SSWG for this end product consists of a review of the disposal or decommissioning plan and a systematic effort to identify, analyze, and control hazards associated with terminating the life cycle for the end product. Techniques could include an OHA or job safety analysis.

Step 33: Close out project
The final step to be taken by the SSWG is to formally close out the project, usually by preparing a short formal report and determining the appropriate repository for project documentation.

REVIEW QUESTIONS

1. What are the steps for an individual initiating a system safety program to take?

2. Discuss the rationale for forming a system safety planning group to develop system safety program policies and guidance.

3. Outline and discuss the role of the safety organization in the system safety effort.

4. Outline the recommended composition of a system safety working group (SSWG), and discuss the knowledge and skills that each member is expected to contribute.

5. What are the three core training courses recommended to support the system safety effort?

6. Who prepares the system safety program plan (SSPP)? Who performs the primary tasks outlined in the plan? Who provides the system safety guidance and policy to develop an SSPP?

7. What are some of the types of data required to support a system safety effort? What are some of the desired characteristics of the data retreival system?

8. Who is responsible for hazard tracking and documentation?

9. Who provides the procurement community with support in the contracting effort required for system safety? What is the nature of this support?

10. What is the purpose of update OHAs, and when are they performed?

PART III

ANALYTICAL AIDS

CHAPTER 10

Analytical Trees

Analytical trees can be used in a variety of ways in the system safety effort. The most common application of analytical trees in current system safety programs is probably the use of fault trees for fault tree analysis (FTA). However, analytical trees can also be used as planning tools, project description documents, status charts, and feeder documents for several hazard analysis techniques (including fault tree analysis). Analytical trees can be multipurpose, life cycle documents and represent one of the most useful tools available to managers, engineers, and safety professionals.

Analytical trees are nothing more than graphic representations (Fig. 10-1). They are literally pictures of a project. Analytical trees use deductive reasoning; that is, they start with a general top event or output event and develop down through the branches to specific input events that must occur in order for the output to be generated.

Trees are called *trees* because, basically, they have a structure that resembles a tree, that is, narrow at the top with a single event symbol and then branching out as the tree is developed.

There are two basic types of analytical tree. The positive tree, or objective tree, which is developed to ensure that a system works properly, and the negative tree, or fault tree, which is generally used for troubleshooting and to investigate system failures (Fig. 10-2).

Positive or objective trees are extremely useful planning tools. In the early stages of a project, they can be used to outline project requirements and list alternatives. As decisions are made, they evolve into graphic checklists and also make excellent status charts and project description documents.

These same positive trees are very useful feeder documents for many types of hazard analysis. For example, the item layout for failure mode and effects analysis worksheets can be taken directly from the positive tree for the project. Positive trees can easily be converted into negative trees for troubleshooting or accident analysis.

Negative or fault trees are excellent troubleshooting tools. They can be used

System Safety for the 21st Century: The Updated and Revised Edition of System Safety 2000,
by Richard A. Stephans
ISBN 0-471-44454-5   Copyright © 2004 John Wiley & Sons, Inc.

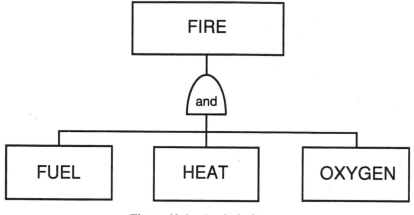

Figure 10-1 Analytical trees.

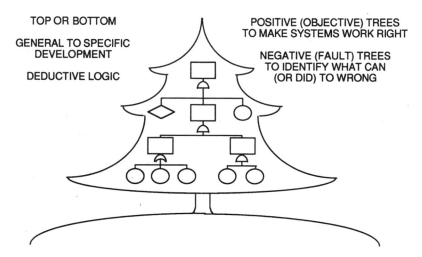

Figure 10-2 Analytical trees. (Derived from the student workbook from the Management Oversight and Risk Tree Workshop presented for the Department of Energy by the System Safety Development Center.)

as preventive tools to identify potential failures before the fact but are more frequently used as accident analysis or investigation tools to pinpoint actual failures.

The management oversight and risk tree (MORT) chart is a large, complex, negative tree (see Chapter 18). Even though the project evaluation tree (PET) is depicted as a positive tree, it is mentally converted and used as a fault tree for accident analysis applications (see Chapter 16).

PURPOSES

Both types of trees have several uses.

1. Analytical trees display clear thinking. By graphically putting a project on paper, the project engineer ensures that he or she has looked at all the alternatives and understands project requirements.
2. Analytical trees force individuals to use the deductive analysis process and to think about the events that must occur at lower levels for output events to be generated.
3. Analytical trees show how relationships and interfaces occur during a project.
4. Root causes are sought and identified when analytical trees are used for accident analysis.
5. Critical paths, that is, the key events or key elements within a particular project, can be determined by the use of analytical trees.
6. Analytical trees make excellent systems status boards. Because the tree is a graphic representation of the project, the current status is readily depicted by color coding the tree or simple checking off completed events or items.
7. Analytical trees also provide a basis for rational decision making, especially early in the concept stage when they characteristically have a number of alternatives and *or* gates.
8. After a project is completed, trees become very good reusable visible records. They provide additional documentation for the project and serve as checklists or guides for repeating similar projects in the future.

TREE CONSTRUCTION

There are seven basic steps in constructing an analytical tree.

1. Define top event.
2. Know the system.
3. Construct the tree.
4. Validate the tree.
5. Evaluate the tree.
6. Study tradeoffs.
7. Consider alternatives and recommend action.

To construct an analytical tree, the first thing to be done is to define the top event. For an objective or positive tree, define what you want to design, build, construct, or accomplish. In the case of a fault or negative tree, identify the type of failure to be investigated.

The second requirement in preparing an analytical tree is to know the system. In and of itself, this requirement is an excellent reason for using analytical trees because it provides reviewers and other parties the assurance that the individual project engineers or others involved with the project do, in fact, know the system well enough put the project on paper logically and systematically in the form of an analytical tree. At times, particularly for consultants, this step may be very time-consuming because it requires considerable research.

The third step is to construct the tree. This step is perhaps the simplest because only a few symbols are involved and the actual construction protocol is straightforward.

The fourth step is to validate the tree, which requires going to another knowledgeable person to have this person (or persons) review the tree for completeness and accuracy.

The next step is to evaluate the tree (and the project) by systematically searching for those places where improvements in the project can be made or where alternatives exist.

The sixth step is to study tradeoffs and further evaluate alternative methods of accomplishing the project.

Finally, consider these alternatives and come up with the recommended action.

Principles of Construction

1. The first of the principles involved in tree construction is to use common symbols. Only a limited number of symbols will be used.
2. Keep it simple. Like any graphic checklist or project outline, it needs to be only as complicated as necessary for those working the project to understand it and be able to follow it. The level of complexity varies, depending on the knowledge level of the participants and, of course, the complexity of the project itself.
3. Keep it logical. Uniform and consistent development from tier to tier and from branch to branch is required. Keep the apples with the apples and the oranges with the oranges.
4. Expect no miracles. An analytical tree as a descriptive document can provide very useful project information and continue to be useful throughout the life cycle of the project, but it is just another piece of paper and, in its simplest form, a graphic checklist. This piece of paper is no better than the effort and knowledge that goes into it and the use to which it is put. Another meaning of "expect no miracles" is to expect things to work in a normal fashion. Do not anticipate a miracle or very remote happenings to save the situation if things go wrong.

5. Use clear, concise titles. Writing inside small event symbols requires concision, if for no other reason, because of the space restrictions.

6. Use "best fit" logic gates. Current thinking is that only two gates are really necessary—the *and* gate and the *or* gate. These gates can be modified with constraint symbols for special situations requiring *priority and* gates, *exclusive or* gates, *inhibit* gates, or *summation* gates.

7. Use transfers carefully. Transfers can be used in two different ways. The tree can be broken into branches and the separate branches put on different pages. These page-break transfers are frequently necessary and do little to inhibit the usefulness or clarity of a particular tree. The other type of transfer, however, needs to be limited. Internal transfers are created where, because of space restrictions or to save drafting effort, one particular branch of the tree that may appear several times is drawn only once and then, through the use of transfer symbols, is transferred to several other places. This practice is very convenient for the draftsperson but difficult for the user, especially if the tree is being used as a status report or as a graphic check list. One of the reasons for analytical trees is to show interrelationships and the relative importance of different parts of the project. A lot of this clarity and perspective is lost by the use of internal transfers. Moreover, the analytical tree then physically has no place to mark things off and to check things off. For example, for an analytical tree to check the air pressure, the condition of the sidewalls, and the condition of the tread on all the tires on a vehicle, that particular branch could be drawn to outline those inspection criteria for one tire and then transfer it to the others. Using the tree to check the vehicle presents problems, however, because it has space for checking off the condition of only one tire (Fig. 10-3). Therefore, use transfers sparingly and attempt to use only page-break transfers, examples of which are shown later in the chapter.

8. Sequence from the left. Many times analytical trees are not specifically concerned about sequencing, but a natural sequence or a sequence that must be followed should be indicated with a constraint symbol, and the events should be sequenced from the left to right.

Analytical Tree Symbols

Analytical tree construction can be accomplished with only seven different symbols, even though some programs like MORT include a few other special purpose symbols.

Three event symbols are used in preparing analytical trees (Fig. 10-4). The general event symbol is a rectangle. The rectangle is the main building block for analytical trees. It is used throughout analytical trees as the event symbol, except in two cases.

A circle is used instead of a rectangle for base events. Base events are

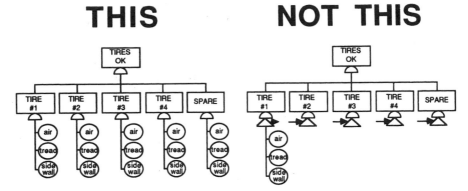

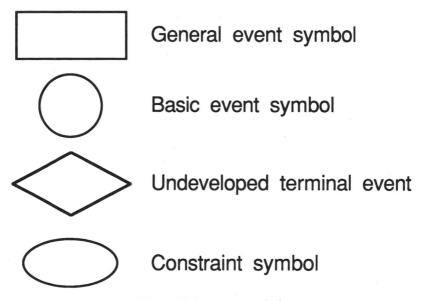

Figure 10-3 Avoid internal transfers. Internal transfers tend to distort the importance of certain branches of the tree and make the tree difficult to use as a status chart or checklist.

Figure 10-4 Event symbols.

bottom-tier events that require no further development or breakdown. There are no gates or events below a base event.

The other event symbol is the diamond, which represents an undeveloped terminal event. The diamond is used in place of a rectangle for a general event that is not going to be further developed (as part of this particular tree at this time). An event may not be developed because the information or the time needed is not available. Sometimes an event is not developed because it is not the responsibility of the individual preparing the tree. For example, most projects require personnel, procedures, and hardware (buildings, tools, equipment, supplies). The personnel manager for the project may logically choose to develop only the personnel part of the tree and show "procedures" and "hardware" as undeveloped terminal events. Like the circle or base event, there are no gates or other events below the undeveloped terminal event symbol.

Another symbol is the oval, which is used as a constraint symbol. Ovals are the footnotes or caveats in the system. They can be used to modify or to show additional information and can be attached to either a general event, a base event, or an undeveloped terminal event. They can be used in conjunction with logic gates. They generally show a specific condition or constraint or modify the meaning of either the event symbol or the gate.

The two gates used in constructing analytical trees are the *and* gate and the *or* gate (Fig. 10-5). The *and* gate is used when, in order to generate an output, having all of the inputs is necessary. For example, in the simple series electrical circuit shown in Figure 10-6, in order for light bulb *D* to be on, switch *A* must be on or closed, switch B must be on or closed, *and* switch *C* must be on or closed.

The other gate is the *or* gate. In order to generate an output with the *or* gate, all that is needed is one or any combination of inputs, as shown by a parallel circuit analogy. In the example in Figure 10-7, for light bulb *D* to be on, switch *A*, switch *B*, switch *C*, switches *A* and *B*, switches *A* and *C*, switches *B* and *C*, *or* switches *A*, *B*, and *C* must be on. In other words, any switch or any combination of switches will, in fact, turn on the light bulb.

In some situations, a simple *and* gate or *or* gate is not adequate. For example, what if a specific combination of inputs is required to generate the

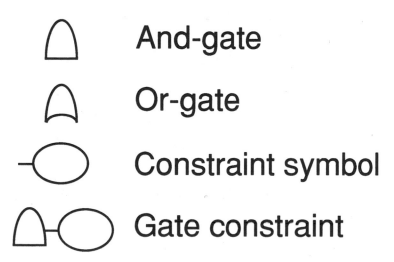

Figure 10-5 Logic gates.

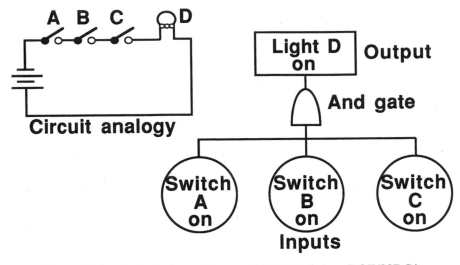

Figure 10-6   And-gate logic. (Source: MORT Workshop, DOE/SSDC.)

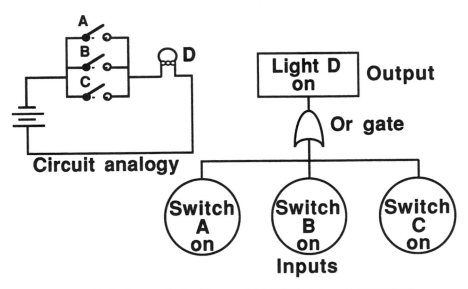

Figure 10-7   Or-gate logic. (Source: MORT Workshop, DOE/SSDC.)

output? What if events must occur in sequence? What if only one input is allowed?

Special situations can be addressed by attaching a constraint symbol to an *and* gate or *or* gate and spelling out any special condition or requirement in the constraint symbol.

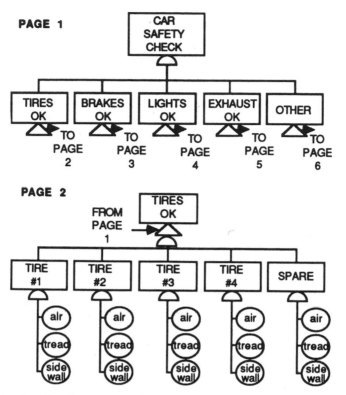

Figure 10-8   Page break transfers. This type of transfer is often necessary for larger trees. Properly done, the clarity and usefulness of the tree is preserved.

The last symbol is the transfer symbol, the triangle. An arrow is commonly used with the triangle to indicate the location to or from which the branch is being transferred (Fig. 10-8). For clarity, the top block of the branch being transferred may be duplicated.

Events and gates are now usually connected with single lines. Some early programs used multiple lines into the input side of gates (Fig. 10-9).

Logically, any general event symbol (rectangle) has one gate attached to it and two or more events as input to that gate. Circles (base events) and diamonds (undeveloped terminal events) are bottom-tier or terminal events that have nothing below them.

The actual physical arrangement of events can be as varied as necessary to take advantage of available space and make the tree as orderly and consistent as practical (Fig. 10-10).

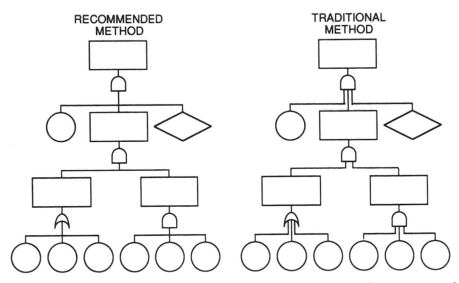

Figure 10-9 Line conventions. Multiple lines are more time consuming to create and tend to clutter the tree.

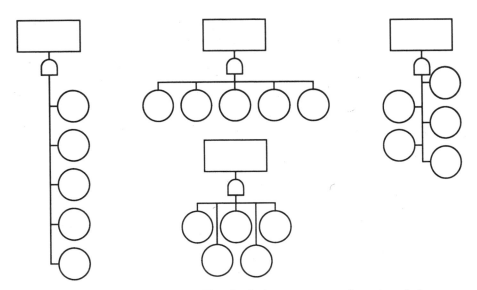

Figure 10-10 Event arrangements. The physical arrangement of event symbols on any given tier is not significant but should be as consistent as practical throughout the tree.

FAULT TREES VERSUS FAULT TREE ANALYSIS

Even though the basic symbols are the same, there are very distinct differences between the fault trees discussed in this chapter and fault tree analysis, the analytical technique discussed in Chapter 15.

Fault trees are excellent troubleshooting tools. Properly used, they can aid in determining what failed or what was less than adequate in a particular system. Fault tree analysis is a tool used to determine how and why the particular failure occurred.

Use of Analytical Trees

A simple example may be helpful in understanding the role of analytical trees as important life cycle, multipurpose documents. The top event or goal, for the example, is to obtain a home computer system (Fig. 10-11).

As with many positive or operational readiness trees, the first tier addresses the personnel, procedures, and plant and hardware that comprise the system. The only personnel for this system is the individual preparing the tree who will be the sole operator of the system: therefore, it is represented by *self* shown as a base event. Procedures for this system consist primarily of the operating and maintenance instructions provided with the hardware and software. Because several sets of instructions may be involved, instructions cannot appropriately be shown as a base event, yet the instructions are not the focus of the exercise. Therefore, they are shown as an undeveloped terminal event. The real focus of the effort is the plant and hardware, in this case, the home computer. The home computer facilities and equipment consist primarily of the computer hardware and the software to make it operate. Other items are also required, like a room to set up the operation, furniture, and electrical outlets, but again they are not the focus of the effort and are therefore shown simply as *other* in an undeveloped terminal event symbol.

In this example, the individual obtaining the home computer system is interested primarily in word processing capability that is compatible with the word processing package at the office. The immediate concern, then, is to obtain system software and word processing software. Other programs that may be added later are not of immediate interest and are again indicated by an undeveloped terminal event. The constraint or requirement for all software is that it is compatible with that used at the individual's workplace.

Most of the decisions to be made in this example involve, then, the selection of the hardware associated with the system. An obvious constraint is that the hardware must be capable of running the desired software. Another constraint, in this particular case, is that the total bill for the hardware can not exceed $2,000. Space restrictions have required the use of page-break transfers for the monitor and the central processing unit (CPU).

Even though not shown, the *monitor* branch would list alternatives such as color versus monochrome, screen size, and manufacturer. Similarly, the CPU selection could evaluate memory size, clock speed, type of processor, output parts, and manufacturer.

The *drives* branch indicates that some decisions have already been made about the drives. The *and* gate indicates that both a hard disk drive and a floppy disk drive are desired. The constraint symbol indicates that internal

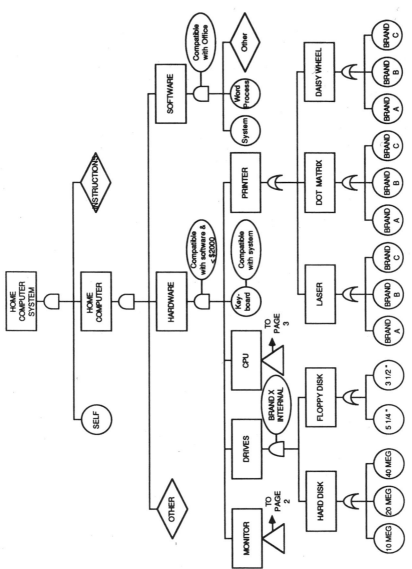

Figure 10-11 Planning tree. Used in the earliest stages of a project, the planning tree will normally contain many options, indicated by the "or" gates.

drives are preferred, and even the brand has been selected. The only remaining decisions are the capacity and size of the drives.

The only requirement this individual has for a keyboard is that the keyboard be compatible with the system. The keyboard was considered a base event.

The final item to be selected in this simplified example is the printer. Three different brands were considered for laser, dot matrix, and daisy wheel printers.

The tree in Figure 10-12 is an example of the type of tree that can be generated as a planning tool for any project. Upper tiers are generally contain *and* gates, but lower tiers list alternatives. This tree is then studied and alternatives are evaluated. As decisions are made, the *or* gates drop out, and the tree evolves into a graphic checklist (Fig. 10-12). This free is now a graphic representation of this home computer system.

During the purchasing and acquisition process, the tree can be used as a status chart (Fig. 10-13). The current status of each item can be depicted on the tree in the manner shown, or the tree can be color coded with highlighters or colored pushpins.

The tree can then be converted into a master fault tree for use as a troubleshooting guide (Fig. 10-14).

A fault tree is simply a mirror image of the positive or objective tree. The objective tree is converted to a fault tree by changing all the *and* gates to *or*

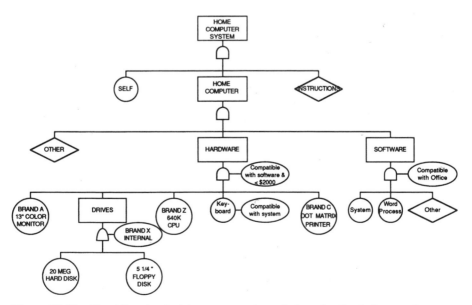

Figure 10-12   Checklist. As decisions are made and plans finalized, the tree becomes a graphic checklist.

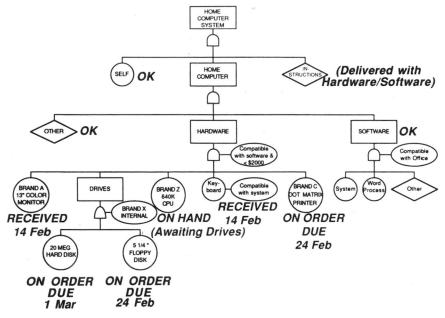

Figure 10-13 Status chart. The tree can be used as a status chart for the project by making notations and/or by color coding.

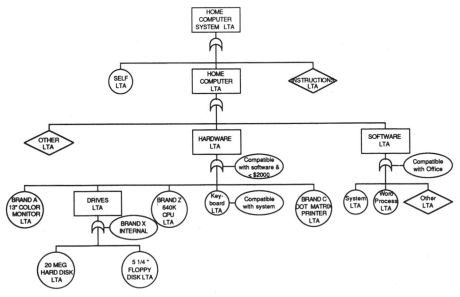

Figure 10-14 Fault tree. The checklist tree can be a troubleshooting guide if the system fails or is less-than-adequate.

gates, by changing any *or* gates to *and* gates, and then by adding *LTA* (less than adequate) to each event symbol.

Benefits of Analytical Trees

Analytical trees make excellent project description documents because they are relatively easy to construct and easy to understand. Many end users and some safety professionals, engineers, and managers are not skilled in reading blueprints, electrical schematics, or other technical documents. The analytical tool is an excellent means for communicating project information within a system safety working group.

Analytical trees are also very useful as feeder documents for several hazard analysis techniques, for example, failure mode and effects analysis (Chapter 14), fault tree analysis (Chapter 15), energy trace and barrier analysis (Chapter 13), and project evaluation tree analysis (Chapter 16), the primary hazard analysis tools for many projects. Virtually any analytical technique or any type of analysis can be simplified by starting with the analytical tree as a base document.

REVIEW EXERCISE

Construct an analytical tree. The tree can be on any subject and either a positive tree or a negative tree. Include at least four tiers in its longest branch, not including the top event as a tier.

REFERENCES

Buys, J.R. 1977. *Standardization Guide for Construction and Use of MORT-Type Analytical Trees.* ERDA 76-45/8; SSDC-8. Idaho Falls, ID: Energy Research and Development Administration.

Clements, P. L. 1987. *Concepts in Risk Management.* Sverdrup Technology, Inc.

Crosetti, P. A. 1982. *Reliability and Fault Tree Analysis Guide.* DOE 76-45/22; SSDC-22, Idaho Falls, ID: Department of Energy.

Hammer, Willie. 1972. *Handbook of System and Product Safety.* Englewood Cliffs, NJ: Prentice-Hall.

Johnson, William G. 1973. *MORT, the Management Oversight and Risk Tree.* Washington, DC: U.S. Atomic Energy Commission.

Johnson, William G. 1980. *MORT Safety Assurance Systems.* New York: Marcel Dekker.

National Aeronautics and Space Administration. 1987. *Methodology for Conduct of NSTS Hazard Analyses.* NSTS 22254. Houston: NSTS Program Office, Johnson Space Center.

CHAPTER 11

Risk Assessment and Risk Acceptance

RISK MANAGEMENT CONCEPTS

Risk management is the general term given to the process of making management decisions about risks that have been identified and analyzed. The traditional risk management choices have been the four *Ts*: terminate, treat, transfer, and tolerate.

Risks can be terminated by either engineering the risk out of the system by making major changes in the types of energy and/or materials being used or by terminating the operation or process with which the risk is associated. Because totally eliminating a risk is usually impossible without eliminating the activity, it may not be a viable choice for private industry and frequently is not a choice for government projects.

Risks can be treated by the application of engineering and administrative controls to reduce the level of risk. It is done by application of systematic application of the system safety precedence (design, safety devices, warning devices, procedures, and training; see Chapter 12).

Risks can also be transferred. The most common way for an organization to transfer risks is by buying insurance. Risk management as a discipline largely involves the study and application of insurance administration. The corporate risk manager normally has a finance-related degree or background and the responsibility for determining the types of insurance, carriers, deductibles, and coverage limits for the organization.

Finally, risks that are acceptable to the organization and/or cannot be terminated, treated, or transferred, are tolerated.

Even though risk management may surface from time to time as the buzzword of the moment in government agencies, because the government is self-insured (or uninsured), the transfer option has little meaning and often the terminate options are also limited. Thus the risk management process for many

System Safety for the 21st Century: The Updated and Revised Edition of System Safety 2000, by Richard A. Stephans
ISBN 0-471-44454-5   Copyright © 2004 John Wiley & Sons, Inc.

government projects consists simply of treating risks until they can be tolerated by establishing risk acceptability criteria, assessing risks, and continuing to apply controls and reassess risks until acceptability criteria are met.

Even though private corporations may have additional options, risk assessment and risk acceptance are important aspects of all risk management programs.

Risk Assessment

One of the major tasks of any systems safety effort is to communicate risk information to management accurately and effectively. Because a number of hazards may be associated with a given project, there must be a way to evaluate these hazards, prioritize them, or quantify them and communicate the risk associated with each to management.

Some definitions are necessary.

Risk: An expression of the possibility of a mishap in terms of hazard severity and hazard probability.

Risk assessment: An evaluation of risk in terms of severity and probability.

Risk acceptance: The process of identifying, quantifying (to the maximum practical degree), and accepting risk by the appropriate level of management.

Controlled risk: The level of risk after controls are applied.

Residual risks: An expression of probable loss from hazards that have not been eliminated by designs.

Risk management: The overall process of identifying, evaluating, controlling or reducing, and accepting risks.

Different systems and approaches attempt to communicate risk information, most using various types of numerical scales. The most widely used is probably the risk assessment code.

Even though this principle is the most widely used, numerous versions of risk assessment codes exist, as pointed out in Chapter 1.

Following is an example of a typical risk assessment effort; it uses the approach proposed by the U.S. Army Corps of Engineers for its facility system safety program.

Facility Risk Assessment. The facility risk assessment effort begins with an initial risk categorization effort made very early in the concept phase of the project. The purpose of risk categorization is to serve as an indicator for the level of effort and scope of the system safety effort.

The risk of the proposed facility is initially categorized as Low, Medium, or High by evaluating such factors as energy sources (types and magnitudes), usage, occupancy, experience (historical data), location, and mission.

Low-risk facilities require a minimal system safety effort and little user involvement.

Medium-risk facilities normally require at least a preliminary hazard analysis and some user involvement.

High-risk facilities require a comprehensive system safety effort and a high level of user involvement.

After the initial risk categorization, the next step in the risk assessment process is the systematic identification of hazards associated with the project. Hazard discovery is the responsibility of all members of the system safety working group and others associated with the project. Other risk management responsibilities are dependent on the level of the risk, with decisions involving more significant risks reserved for senior commanders or managers.

The hazard discovery process, which is the foundation of the risk assessment effort, continues throughout the project and relies on input from many sources, including

PHL

PHA

User-defined unacceptable or undesirable events

Design reviews

Hazard analysis outputs

Change order reviews

Health hazard assessment reports

Individuals who discover a hazard

Formal hazard discovery forms (Fig. 11-1) may be used to ensure that hazards that are identified are documented, evaluated, and entered into a hazard tracking system.

Hazard information is converted to risk information by evaluating the severity of potential accidents associated with the hazard and by evaluating the probability that the hazard could produce an accident. It is done by developing a matrix with severity on one axis and probability on the other, with a numeric code used to represent the risk associated with each hazard. This risk assessment code (RAC) is used to prioritize hazards and determine their acceptability. Hazard severity may be expressed quantitatively (for example, dollar loss or number of injuries), qualitatively (verbal descriptions), or as a combination (Table 11-1).

Similarly, mishap probability can be expressed with numerical probabilities or word descriptions (Table 11-2).

The risk assessment code matrix shown in Figure 11-2 is typical of those used in military programs and very similar to the examples found in MIL-STD-882B.

An element of the facility system safety effort not found in MIL-STD-882B and not typical of other programs is the control rating code (CRC).

HAZARD DISCOVERY FORM

Part I Discovery

1. Facility/Item Identification:
     a. Submitted by: _____ Date: _____
        Submitter's Organization: _____
     b. MILCON Project Number: _____ Facility ID: _____

2. For each hazard identified, provide the following data:
     a. Hazard Description: _____
     _____
     b. Potential Consequences: _____
     _____
     c. Submitter's Proposed Severity _____
        Probability _____
     d. Identify any applicable source documents and/or references:
     _____
     e. Recommended actions to reduce the risk (optional) _____
     _____
     _____

HAZARD DISCOVERY FORM
(continued)

Part II Review

     a. Reviewed by: _____ Date: _____
     b. Reviewer's RAC: _____
     c. Reviewer's Recommendation(s): _____
     _____
     d. Document actions taken, as applicable, to include entry into Hazard
        Tracking Log and identification of project documents affected or
        modified as a result of this action. Also, document any decision to
        reject the submitter's recommendations.
     _____
     _____
     _____

Part III Approval of Proposed Actions

     Approved _____ Date: _____

     Copies Furnished _____

Figure 11-1   (top) Hazard discovery form, Part I; (bottom) Parts II and III. (Source: U.S. Army Facility System Safety Manual. Draft.)

TABLE 11-1. Hazard Severity

I Catastrophic	II Critical	III Marginal	IV Negligible
May cause death or loss of a facility	May cause severe injury, severe occupational illness, or major property damage	May cause minor injury or minor occupational illness resulting in lost workday(s) or minor property damage	Probably would not affect personnel safety or health and thus less than a lost workday but nevertheless is in violation of specific criteria

TABLE 11-2. Mishap Probability

A Frequent	B Probable	C Occasional	D Remote	E Improbable
Likely to occur frequently	Will occur several times in life of a facility	Likely to occur sometime in life of a facility	Unlikely but possible to occur in life of a facility	So unlikely that it can be assumed that occurrence may not be experienced

The CRC concept, believed to have been pioneered by Ludwig Benner of Events Analysis, Inc., applies RAC logic to hazard controls to develop a matrix to evaluate alternative control measures. It is used for selection and evaluation of hazard controls in conjunction with RACs. It is based on system safety precedence. The rules for use are

The CRC should never be more than one digit higher than the RAC it controls.

No single-point failures are allowed in severity level I and II mishap

RAC 1 and 2 hazards cannot be controlled with cautions, warnings, or personal protective equipment only.

The energy controls that form one side of the CRC matrix roughly parallel and are consistent with the system safety precedence.

Eliminate energy sources

Limit energy accumulated

Prevent energy release

Probability	Severity			
	I Catastrophic	II Critical	III Marginal	IV Negligible
A Frequent	1	1	1	3
B Probable	1	1	2	3
C Occasional	1	2	3	4
D Remote	2	2	3	4
E Improbable	3	3	3	4

Figure 11-2 Risk assessment code matrix used in the facility system safety effort. It is typical of military programs and similar to the example matrices listed in MIL-STD-882B. (Source: U.S. Army Facility System Safety Manual. Draft.)

Provide barriers to energy flows on the energy source, between the source and the target, and/or on the target

Change the energy release patterns by reducing the transfer and/or release rate, changing the release distribution, and/or relocation energy or targets

Minimize harm by hardening vulnerable targets

Treat harm

The other axis of the matrix (Fig. 11-3) evaluates the manner in which the energy control is integrated into the system (for example, design, passive device that provides protection automatically, active safety device that requires application or activation by the user or operator, or warning device only).

Facility Risk Acceptance. The following risk acceptability criteria are typical of those in MIL-STD-882B:

RAC 1—Unacceptable
RAC 2—Undesirable
RAC 3—Acceptable with controls
RAC 4—Acceptable

		Design	Passive Safety Device	Active Safety Device	Warning Device
		I	II	III	IV
A	Eliminate Energy Source	1	1	2	3
B	Limit Energy Accumulated	1	1	2	3
C	Prevent Release	1	2	2	3
D	Provide Barriers	2	2	3	4
E	Change Release Patterns	2	3	4	4
F	Minimize/Treat Harm	3	3	4	4

Figure 11-3   Control rating code for CRC matrix. (Source: U.S. Army Facility System Safety Manual. Draft.)

The facility system safety risk assessment process works such that if the initial risk assessment produces a RAC of 1 or 2, some type of control must be applied. After this control is applied, a second risk assessment or "controlled RAC" is determined. First, the control rating code is evaluated to ensure that the CRC rules have been met and then to ensure that the controlled RAC is 3 or 4. If the CRC rules are met and the controlled RAC is 3 or 4, the corrective action is taken and the risk has been reduced to an acceptable level. However, if the CRC rules have not been met and/or the RAC remains at 1 or 2, other controls must be applied and the reassessment cycle is repeated. If other controls are not available, then risk acceptance decisions must be made by the appropriate level of management, and risk acceptance decisions must be documented (Fig. 11-4).

Residual risks should be formally accepted and documented as outlined in the acceptance criteria and the system safety program plan (Fig. 11-5).

The RAC matrix and the risk acceptability criteria described in this example are representative of many of the risk assessment efforts presently in use and are consistent with the MIL-STD-882B approach to risk assessment.

A risk management program consisting primarily of hazard discovery and evaluation, initial risk assessment, comparison with acceptability criteria, application of controls to reduce the risk assessment to acceptable levels (repeated as necessary), and formal acceptance of residual risks is also representative of government programs.

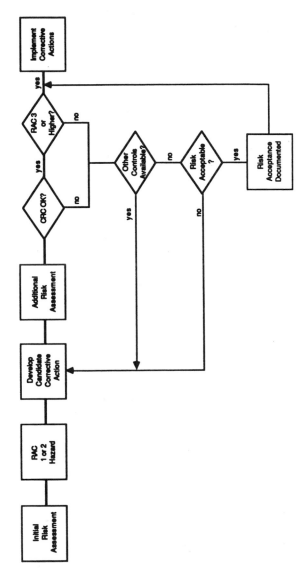

Figure 11-4 Recommended RAC 1 and RAC 2 risk assessment process. This flow diagram outlines the process for accepting risks associated with hazards that have initial risk assessments that are unacceptable or undesirable. (Source: U.S. Army Facility System Safety Manual. Draft.)

RESIDUAL RISK ACCEPTANCE FORMAT
Part I Description of Residual Risk

1. Facility/Item Identification:

   MILCON Project Number: _____ Facility ID: _____

2. For each proposed acceptance of a risk (e.g., a less than optimum RAC or CRC) associated with an identified hazard, provide the following:

   a. Hazard description and potential consequences: _____
   _____
   _____

   b. Final Risk Assessment Code: _____

   c. Final Control Rating Code: _____

   d. Identify source document(s) and/or reference(s): _____
   _____

   e. Document any alternative actions to reduce the risk: _____
   _____

   f. Proposed By: _____ Organization: _____
      Date: _____

3. Technical Review By Appropriate Level Based on RAC.

   a. Reviewed By: _____ Date: _____

   b. Reviewer's Recommendations: _____
   _____

Part II Approval

   Signature: _____ Date: _____

Figure 11-5 Residual risk acceptance form to be used for formal acceptance of residual risks. (Source: U.S. Army Facility System Safety Manual. Draft.)

The initial risk categorization process and the use of control rating codes are not addressed in MIL-STD-882B and are believed to be unique to the facility system safety effort.

As discussed in Chapter 4, current risk assessment efforts show some shortcomings.

RISK ASSESSMENT SHORTCOMINGS

As pointed out in Chapter 4, deficiencies in the current risk assessment code systems represent one of the major problems facing the entire system safety effort.

The quantitative severity and probability scales used in most RAC matrices are very subjective. The top of the severity scale is usually a dollar loss of about $500,000 or a single fatality, even though many hazards are capable of producing hundreds or thousands of fatalities and hundreds of millions of dollars in financial loss in a single accident.

For all practical purposes, the risk assessment code is the driver in the system. The value of this code prioritizes the fixes and the management or command emphasis given to a particular problem. If these codes are not detailed enough to discriminate between very serious hazards and lesser hazards accurately, they are of little value.

TOTAL RISK EXPOSURE CODES

Following is a possible alternative or complement to the current risk assessment code system. The most significant differences in the total risk exposure code and the current risk assessment code are that the range of the severity scale is greatly expanded, the magnitudes of all losses are converted to dollars, and an "exposure" scale is substituted for the "probability" scale as the second axis for the matrix. The exposure for each hazard is determined by multiplying the probability of a single occurrence (expressed in number of occurrences per 100,000 exposure hours) by the total estimated exposure hours during the life of the system and by the number of systems to be produced.

The ten severity codes (Table 11-3) represent single event losses from less than $100 to over $10 billion, with each severity code increasing by an order of magnitude. This scale provides a more meaningful assessment of the hazards associated with systems capable of producing multiple deaths and hundreds of millions or even billions of dollars in total losses (such as transportation system accidents).

Ten exposure codes (Table 11-4) represent estimates of the total number of accidents for the system. The low end of the scale (exposure code of 1) rep-

TABLE 11-3. Severity Codes

Code	Range	Average
10	>10 Bil	5×10^{10}
9	1–10 Bil	5×10^{9}
8	100 Mil–1 Bil	5×10^{8}
7	10–100 Mil	5×10^{7}
6	1–10 Mil	5×10^{6}
5	100 K–1 Mil	5×10^{5}
4	10–100 K	5×10^{4}
3	1–10 K	5×10^{3}
2	100–1000	5×10^{2}
1	<100	5×10^{1}

TABLE 11-4. Exposure Codes

Code	Range	Average
10	>1000	5×10^3
9	100–1000	5×10^2
8	10–100	5×10^1
7	1–10	5×10^0
6	0.1–1	5×10^{-1}
5	.01–0.1	5×10^{-2}
4	.001–.01	5×10^{-3}
3	.0001–.001	5×10^{-4}
2	.00001–.0001	5×10^{-5}
1	<.00001	5×10^{-6}

EXPOSURE CODE

SEVERITY CODE	10	9	8	7	6	5	4	3	2	1
10	20	19	18	17	16	15	14	13	12	11
9	19	18	17	16	15	14	13	12	11	10
8	18	17	16	15	14	13	12	11	10	9
7	17	16	15	14	13	12	11	10	9	8
6	16	15	14	13	12	11	10	9	8	7
5	15	14	13	12	11	10	9	8	7	6
4	14	13	12	11	10	9	8	7	6	5
3	13	12	11	10	9	8	7	6	5	4
2	12	11	10	9	8	7	6	5	4	3
1	11	10	9	8	7	6	5	4	3	2

Figure 11-6 Total risk exposure code (TREC) matrix.

resents an estimate of less than one chance in 100,000 that an accident of the given magnitude will result from the hazard during the entire life of the system. The high end of the scale represents an estimate that more than 1,000 accidents will occur as a result of the hazard over the life of the system. Each exposure code increase represents an order of magnitude increase in the estimated likelihood of an accident or accidents occurring as a result of the identified hazard. Even though the unit of time suggested for exposure calculations is 100,000 exposure hours, any units can be used (1,000,000 exposure hours, years, or number of cycles, for example), as long as they are used consistently because these units cancel out of the equation.

The total risk exposure code is then determined by adding the severity code and the exposure code (Fig. 11-6).

The TREC has a real relationship to the estimated dollar losses associated with a system over its lifetime.

Use of the TREC allows the system safety working group to provide the project manager with the following types of information:

Total risk exposure (TRE): The total number of dollars estimated to be at risk as a result of the particular hazard being evaluated. The Total Risk Exposure is calculated quite simply by subtracting 5 from the TREC and adding that number of zeros to 5.

$$TRE = 5 \times 10^{(TREC-5)}$$

For example, a hazard with a TREC of 11 would be expected to produce approximately \$5,000,000 in losses over the life of the project. Controls that reduce the TREC to 9 would then be expected to save an estimated \$4,950,000 (\$5,000,000–\$50,000).

Annual risk exposure (ARE): The annual risk exposure is the total risk exposure divided by the estimate project life in years.

$$ARE = \frac{TRE}{Project\ life}$$

Unit risk exposure (URE): The unit risk exposure is a measure of the projected dollar loss per unit (or prorated per unit) and is calculated by dividing the TRE by the total number of units produced.

$$URE = \frac{TRE}{Number\ of\ units}$$

Risk exposure ratio (RER): The ratio of the risk exposure to the project cost, the RER is calculated by dividing the TRE by the total project budget or projected cost. Assuming a uniform budget and unit cost based on total cost divided by total number of units, the RER can also be calculated by dividing the ARE by the annual project budget or cost or by dividing the URE by the unit cost.

$$RER = \frac{TRE}{Total\ budget}$$

Meaningful risk acceptability criteria could be developed using these terms and/or the TREC.

Like the RAC, the quality of the TREC is dependent on the quality of the input data. A very important system safety task is validating and updating the initial estimates of failure rates, exposure rates, project life, and number of units in the system.

In any event, this information should be more meaningful, more useful, and more valuable to management than the traditional risk assessment code.

REVIEW QUESTIONS

1. Explain how and why risk management is different in government agencies and projects than it is in private industry.
2. Discuss the four *T*s of risk management.
3. Define and discuss risk assessment codes.
4. Define and discuss control rating codes.
5. Define and discuss total risk exposure codes.
6. Discuss the relationship between TRECs and estimated dollar losses.
7. If an identified hazard is capable of causing 250 fatalities and causing $50 million in property damage and other losses, what is its severity code? If this hazard is associated with certain model of aircraft with an expected service life of 25 years, what is the exposure code, given that the fleet has 200 aircraft and the hazard is estimated to be likely to result in an accident no more than once per 5 million flying hours (assume 2,000 flying hours per aircraft annually)? What is the TREC? What would be the estimated losses associated with this hazard over the life of the system?

REFERENCES

Browning, R. L. 1980. *The Loss Rate Concept in Safety Engineering*. New York: Marcel Dekker.

Clements, P. L. 1987. *Concepts in Risk Management*. Sverdrup Technology, Inc.

Hammer, Willie. 1972. *Handbook of System and Product Safety*. Englewood Cliffs, NJ: Prentice-Hall.

U.S. Army. n.d. *Facility System Safety Manual* Washington, DC: HQ Safety, USACE. Draft.

U.S. Department of Defense. 1984 (updated by Notice 1, 1987). *System Safety Program Requirements*. MIL-STD-882B. Washington, DC: Department of Defense.

Human Factors

HUMAN RELIABILITY

A great deal of progress has been made toward improving and evaluating the reliability of hardware systems; however, the place where systems most frequently fail is in the interface of humans with the system. Human reliability is generally much lower and more difficult to control than hardware reliability.

If a stimulus is provided to a piece of hardware, the same response is obtained from that piece of hardware day after day, year after year, until it fails. Even when hardware fails, the failure modes, rates, frequencies, and effects can be predicted with a reasonable degree of accuracy.

People, however, are much more difficult to analyze. A given stimulus provided to a person or group of people may produce entirely different responses over time because of the mediation process that takes place (Fig. 12-1). Individuals judge the input stimulus based upon highly individualized value systems. For this reason, accurately determining what type of response will result from a given stimulus is very difficult. Not only does this response vary from group to group, and from individual to individual but it also varies with the same individual over time. A response today may be totally different from the response to the same stimulus three days from now.

Stress Resistance and Failure Modes

Humans also vary widely in stress resistance, both physiological and psychological. If a specific amount of force is applied in a given direction, at a given point, to a structural component of a building, the result of the stress applied can be determined quite accurately. How much the particular structural member will yield or bend and the point at which it will break can be accurately predicted. However, the stress resistance of humans again varies significantly, not only between groups of individuals but also between individuals over time.

System Safety for the 21ˢᵗ Century: The Updated and Revised Edition of System Safety 2000, by Richard A. Stephans
ISBN 0-471-44454-5   Copyright © 2004 John Wiley & Sons, Inc.

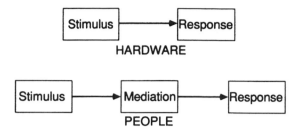

Figure 12-1   Basic response model. (Source: System Safety Development Center (SSDC) course handouts.)

Studies of military personnel indicate that the best peacetime leader or soldier may not be the best in combat. Some seem to function relatively well under the extremely high stress of combat while others who are good performers in a peacetime or administrative environment are very poor performers under high stress.

The ability to perform when ill or fatigued varies widely. Some individuals' performances degrade gradually over time as fatigue or illness sets in. Others may perform at a constant level up to a given point in time, when performance falls off very, very rapidly.

The failure modes of people vary considerably more than the failure modes of hardware. Motivation is a primary factor in determining the level of human performance.

Human performance is a function of aptitude (the basic ability to do something) plus training (formal education, on-the-job training, and experience). These two factors combined determine the overall ability of an individual to perform a particular task. These two factors are then multiplied by motivation.

$$\text{Performance} = (\text{Aptitude} + \text{Training}) \times \text{Motivation}$$

Highly motivated individuals with limited aptitude may produce a higher level of performance than people who have a great deal of talent, skill, and education but little motivation.

A second failure mode for humans that does not apply to equipment is boredom versus vigilance. People are very poor at performing tasks that are boring or require a great deal of concentration, especially in situations with little stimulus or feedback. In contrast, machines can perform boring and tedious jobs over long periods with high reliability.

A third failure mode of humans is related to expectations and stereotypes. People learn to expect certain things to operate in certain ways. When things do not go as expected, human error rates go up significantly. For example, bolts are supposed to go in and nuts are supposed to go on bolts with clockwise rotation. Clockwise rotation tightens; counterclockwise rotation loosens. People learn this stereotype.

A few years ago a major automotive manufacturer decided to put left-hand threads on one side of automobiles so that the motion of the wheels would always tend to tighten the lug nuts. From a purely engineering viewpoint, this decision was sound. Unfortunately, it violated stereotypes and resulted in a lot of groaning and stripped bolts and overtightened lug nuts on one side of the car. In addition to learned stereotypes, there are also natural stereotypes. The most troublesome natural stereotype in the design world and in the world of human factors or ergonomics is left-handed people. The world is designed for right-handed people, and in many cases problems are created and higher error rates induced in the left-handed people.

A fourth failure mode for humans is perception. People can literally "see" things in different ways. For example, a soft drink can viewed from the end as a circle, viewed from the side as a rectangle, and viewed in perspective is, of course, a cylinder. People also view things differently based on their own individual backgrounds and value systems, which filter the way the world looks. Attitudes toward safety and toward rules, regulations, and procedures are dependent upon how they are perceived. For example, most workers seem to accept wearing personal protective equipment as proper and normal and would feel uncomfortable performing work tasks without it. Some, however, perceive personal protective equipment as sissy stuff and seem to feel that wearing it is not consistent with their macho image.

A further human failure mode, known as *coupling*, is a situation in which two really unrelated types of situations, events, or conditions are linked mentally or emotionally.

A few years ago I gave a MORT seminar in Las Vegas. When one of the attendees showed up at the hotel, a mistake with his reservations led the hotel to give him a very low rate and a very nice corner suite. The next night he dropped a few coins into a nickel slot machine and hit an $1,150 jackpot. Needless to say, this individual thought this course was the best he had ever attended! The experience of winning money, living well, and having an enjoyable time outside the classroom was coupled with his coursework and significantly influenced his attitude toward it.

The variability in stress resistance and the multiple failure modes of humans tend to make human error rates relatively high.

HUMAN ERROR RATES

Figure 12-2 depicts error rates for a variety of tasks. Unfortunately, the points depicted represent averages. The amount of variance or the ranges of performance are not included, yet the information indicates the impact of stress on human performance.

The scale at the top of Figure 12-2 represents the frequency of human

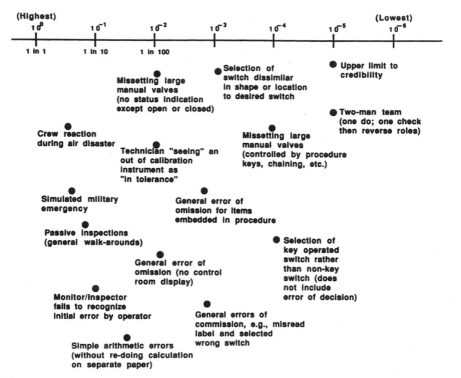

Figure 12-2 Estimates of human performance error rates. (Source: SSDC course handouts. Data points were compiled from various government studies.)

errors. The left side of the scale represents the highest error rates for the situation in which an error would be made every time a task is performed and progresses to the situation on the right side of the scale, where a human error would be expected only one time for every million times a task is performed. Note, however, that the upper limit of credibility for human performance is about one error for every 100,000 tries. This level of performance is accomplished by using a two-person team with one person performing the task while being monitored by the other and then reversing roles. This type of system is used for operations involving nuclear weapons and other critical systems. The more interesting and disturbing information on the chart is found on the other end of the scale. It is difficult to interpolate with precision, but about one time in five apparently a crew reaction during an air disaster is going to be done improperly. A classic example of this type of error involves twin-engine aircraft. The correct reaction to many in-flight engine problems is to shut down the malfunctioning engine. Not uncommonly, the pilot shuts down the good engine instead of the malfunctioning engine. This reaction occurs in all types of twin-engine aircraft, including commercial airliners, for example, the 737 crash in England, January 1989. The aircrews may be experienced, highly skilled, well-trained professionals, and motivation should be quite high in such situations, yet the error rates remain surprisingly high.

The implication is that systems that rely on correct aircrew performance of critical operations during in-flight emergencies have relatively high failure rates built into them. To provide safe twin-engine aircraft, design engineers must not only design aircraft that will fly on one engine but also provide improved means of correctly identifying the malfunctioning engine and/or engineering controls to prevent pilots from shutting down the wrong engine when an engine fails.

The error rate also appears to be quite high (perhaps 20%) during simulated military emergencies. If error rates approach this level in training exercises, imagine what they must be in actual combat situations. Even though accurate statistics are limited (safety recordkeeping has not had a high priority in most wars), available information reveals that accidents have historically accounted for a very significant percentage of total losses.

The failed Iranian hostage rescue effort, the losses suffered on the USS *Stark*, the inadvertent shooting down of a commercial airliner during naval operations in the Persian Gulf, and the rash of U.S. Navy accidents during 1989 and 1990 all indicate the tremendous potential for loss when sophisticated weapons systems are handled by individuals under significant psychological and/or physiological strain.

The safety community should note that, according to this data, a general walk-around inspection will fail to identify about 10% of workplace hazards.

The real message is that, especially in high-stress situations, humans are the least reliable component in many systems.

IMPROVING HUMAN RELIABILITY

If people really are the weak links in many systems, then some significant safety improvements can be made by addressing human factors in system design.

For years, the importance of a good quality assurance (QA) or quality control program for critical hardware has been recognized. Some type of QA effort for people with critical duties is probably equally important. Elements of such programs are already in place in the traditional high-risk areas that created the system safety effort, specifically the aerospace and nuclear communities. Existing programs tend to concentrate on initial selection criteria and standards, education and training requirements, and physical condition. All of these factors are important; however, many other factors can also significantly affect human reliability. As previously noted, psychological stress greatly affects human performance. This area is difficult to monitor because doing this job properly involves getting into areas that we traditionally and correctly feel are part of a person's private life. For example, marital problems, drug or alcohol habits, financial difficulties, and problems with kids can all have a very significant impact upon human performance, yet in these areas monitoring is difficult and, by and large, inappropriate. Such monitoring can be an invasion of privacy and a violation of civil liberties. At the same time, in those areas where human error can be catastrophic, there is, in fact, an obligation to provide a safe workplace and ensure public safety, and key individuals may have to sacrifice a part of their privacy. Coupled with programs to monitor the level of stress should be employee assistance programs to help people cope with and reduce the stress related to personal problems. An expanded effort ensuring that individual physical and mental capabilities, education and training, and motivation are properly matched with job and task requirements can aid in improving the overall reliability of most systems.

The frequency of human error can also be reduced by better design of input and output systems. The design engineer with a knowledge and understanding of both the physical and mental limitations of system users can reduce operator errors by applying sound ergonomic principles to the design of machine controls, switches, keyboards, and other input devices. Similarly, instruments, gauges, and various other types of readouts or system outputs that carefully consider the capabilities and limitations of the operators and output monitors can enhance the communication between the human and mechanical components of the system.

People perform better if parameters such as temperature, humidity, light levels, noise levels, and other environmental factors are maintained within comfort limits. Air conditioning, good illumination, soundproofing, and other creature comfort measures tend to reduce both psychological and physiological stresses and thus allow humans to work at higher levels of performance for longer periods. These measures are probably more important, albeit more difficult to implement, in combat vehicles and construction equipment than they are in administrative settings.

Because the ability to change basic human characteristics is rather limited, the most significant improvements in overall system reliability may be made by reducing the role of the human, especially in those situations where human performance is most error prone. The development of artificial intelligence and sophisticated, computerized control systems offer tremendous potential for controlling human errors by the skillful integration of man and machine. In situations and environments where human error rates may not be acceptable, the primary role of humans may well be to provide system oversight and monitoring of key automated functions and work in parallel with automated control systems. Whereas basic human capabilities tend to remain relatively fixed, machine capabilities continue to expand at very rapid rates.

In summary, the design engineer and the safety engineer must look at total system design. Designers, aided by system safety working groups, must address the total system to ensure that systems are less likely to induce human errors and are more tolerant of human errors. Critical systems should be capable of tolerating single-point failures of human or hardware components.

System Safety Precedence

Another means of improving overall system reliability is by application of the system safety precedence (or safety precedence sequence or hazard reduction precedence sequence).

The first step is to design for minimum hazard. It includes the use of the safest possible energy sources and forms of energy and careful design of the hardware to ensure that the hardware itself will not fail at an unacceptable rate, that the failures that do occur are not catastrophic and do not produce unacceptable results, and that the hardware design tends to be human error reducing and tolerant rather than error inducing.

After all practical steps have been taken to eliminate unwanted energy flows and harmful environmental conditions by design and engineering, the next step is to install physical barriers to protect potential targets from any unwanted energy flows or harmful environments. Physical guards and barriers properly designed, installed, and maintained can provide positive protection from many hazards. Passive guards and barriers tend to be more reliable than active devices because they eliminate the need for human action. For example, automotive airbags are passive devices, whereas seatbelts are active devices. Walls, fences, barricades, machine guards, sprinkler systems, lightning rods, ground fault circuit interrupters, interlocks, pullback devices, and pressure relief valves are all examples of safety devices designed to block, channel, restrict, or otherwise control unwanted energy flows or harmful environments.

Personal protective equipment is sometimes included in this category, although generally it provides marginal protection in comparison to more substantial passive barriers. It normally represents the last line of defense, and its use and effectiveness may rely on procedures and training.

After all practical guards, barriers, and other safety devices have been integrated into the system, the next items to consider are warning devices.

Design and engineering controls and physical guards and barriers may provide positive protection and actually eliminate the occurrence of certain types of accidents.

Warning devices do not provide positive protection but can aid in preventing accidents and/or reducing the consequences of unwanted energy flows. Some warning devices can alert operators and others in time to prevent an unwanted energy flow; others simply provide time to escape. Fire alarms, air quality monitors, radiation alarms, backup alarms, and annunciator panels are all examples of devices that warn of impending danger but do nothing, in and of themselves, to prevent or control the hazard.

The lowest level of control relies on procedures and training to recognize and avoid workplace hazards. Even though procedures and training are the least effective control, they are probably the most frequently used and the quickest and cheapest to implement.

In actual practice, all of these controls are frequently used in combination. In performing system safety analyses, accident investigations, and design

reviews of safety engineering, however, good practice requires systematically using the system safety precedence for selecting and evaluating hazard
controls.

HUMAN FACTORS FOR ENGINEERING DESIGN

In attempting to integrate human factors into design effort and to evaluate
the adequacy of designs, four specific areas are analyzed.

Behavioral Stereotypes

The first area of concern is violation of behavioral stereotypes. As previously
noted, when people expect things to operate one way and they operate differently, high error rates can be expected. Certain types of behavior are natural
for any given individual. When an individual crosses his or her arms, the left
hand may be on top, the right hand may be on top, or neither hand may be on
top, but an individual naturally crosses his or her arms the same way every
time for a lifetime. Similarly, when hands are crossed, the same thumb is naturally always placed on top. Whether the left thumb or the right thumb goes
on top is not necessarily related to whether the individual is left-handed or
right-handed. Both of these are natural stereotypes, and in both cases trying
to perform even these simple tasks in a manner that violates the stereotype is
difficult or awkward. Fortunately, neither of these actions is normally required
to operate equipment or perform other tasks and thus do not present significant problems. The most significant natural stereotype is which hand a person
uses naturally. Many devices designed for right-handed people induce high
error rates or reduce effectiveness or efficiency when used by left-handed
people. Learned behavioral stereotypes are more of a concern than natural
stereotypes, however, because there are many more of them and because they
can vary from group to group and from time to time.

   As previously mentioned, the common or expected way to tighten a bolt or
screw is to turn it clockwise, which is also the "normal" way to close a valve.
For most toggle switches, up is on and down is off. Red usually means stop or
danger, and green indicates on or OK. These conventions are fairly universal.
Many learned behavioral stereotypes vary from group to group. For example,
those of us who learned to drive in the United States or most parts of Europe
find driving on the right side of the road natural. Driving on the right side of
the road is "unnatural" (and very dangerous) in Japan, the United Kingdom,
and various other parts of the world.

   The automotive and aviation industries provide many examples of error-
inducing designs. Most people tend to look for a car horn near the center of
the steering wheel, not on the end of the turn signal lever. Designers of high-
performance aircraft with movable sweep wings felt that moving the wings to
the forward position by moving a control forward and moving them to the

swept or rearward position by moving them rearward was perfectly logical. However, the wings are moved forward when the aircraft slows down for landing and are moved rearward when the aircraft accelerates for high-speed flight. The throttle control thus moves forward for the aircraft to go faster and rearward for the aircraft to go slower. Thus is it more "natural" for the wing controls to move in the same direction as the throttle or for them to move in the same direction as the wings? No universal agreement yet exists on this question. A popular computer game uses the forward arrow on the control device to raise the nose of the simulated aircraft and the rearward arrow to move the nose of the aircraft downward, but a pilot moves the stick of an aircraft rearward to raise the nose.

In situations with a widely accepted standard, convention, or consensus, a deviation from the expected is likely to produce widespread errors.

A lack of standard layout or design can also be confusing. In mobile truck cranes, for example, pulling back on a given lever on a crane from one manufacturer may raise the load. Pulling back on a similar crane from a different manufacturer may raise the boom, and pulling back on a similar lever from a third manufacturer may swing the boom. Errors should be expected from operators who move from one make of crane to the other.

In any event, designers and those involved in design review should be aware of user and operator stereotypes, ensure that they are not violated, and take extra care to describe the correct movement of controls or other situations in which confusion could occur, especially if incorrect actions could produce catastrophic results. The violation of stereotype is particularly important for actions that must be taken rapidly under emergency conditions because people tend to revert to stereotype in such situations, that is, take the action that seems most "natural."

This problem alone may justify the formation of a system safety working group that includes the user community.

The problem of behavioral stereotypes emphasizs the need for the designer to ensure that systems fit the users or operators mentally. Systems must also fit the users or operators physically. These objectives are the focus of human engineering or ergonomics.

Human Engineering

This area of human factors is probably the most developed, primarily because it tends to address physical dimensions and hardware. The ergonomics effort is directed at ensuring that control panels are designed in such a way that instruments and gauges are of the appropriate size, type, and arrangement to be easily read and correctly interpreted; that physical dimensions of furniture, equipment, and tools properly match the relevant physical dimensions of users and operators; and that the movement ranges, directions, and locations of control levers, valves, switches, and other apparatus match the natural anatomical functioning of the operators.

The traditional engineering approach was to design for the "average man." The problem with the approach was that virtually no one was an "average man" in all dimensions and that thus few systems matched anyone perfectly. This problem was amplified with the influx of women into the workplace, especially into "traditionally male" occupations, the tools and equipment for which had been designed for average male dimensions. The "average man" in the workplace is now about 50% female!

This problem is a primary area of study for the human factors engineer. Many design manuals and books are available to assist in this aspect of human factors, and much progress has been made in this area.

Man-Machine Interface and Tradeoffs

The concern in this area is not the physical interface addressed by ergonomics but the functional interface. This area of human factors engineering is one of the more difficult because of the rapidly expanding capabilities of control systems using artificial intelligence and because human egos and emotions and even ethical questions can become intertwined in the decision-making process.

Technically, allowing computers to make life-or-death decisions and to take control of complex systems in emergency situations may be appropriate. As previously noted, human error rates tend to be high in these situations, whereas well-designed, properly functioning automated control systems are capable of performing at very high levels of reliability. The automated controls, allowed to function without human interference, would have avoided most of the damage at Three Mile Island. Yet, many of us are reluctant to place our lives in the hands of a machine. If an automated control system malfunctions, the potential for disaster may be many times greater than the damage that would have occurred with a human operator.

Knowing how to ensure that machines do those things that machines do best and that humans do what humans do best and vesting critical decision-making and override authority properly may be the greatest challenges facing the human factors, safety, and design communities.

Misuse Analysis

Misuse analysis examines the corollary to Murphy's Law that states "What can be misused, will be misused." Product liability litigation has been a major driving force in misuse analysis because many times a manufacturer may be held liable for damages resulting from purchasers using or attempting to use products for other than their intended purpose.

The design engineer, aided by the system safety working group, systematically reviews all aspects of systems to determine why and how persons may use the system, subsystem, or component improperly and attempts to identify, analyze, and control hazards associated with misuse.

The age, education level, and other qualifications and characteristics of the potential users are very significant factors to consider in performing misuse analysis. That is the reason children's toys have been the subject of much litigation and significant design improvements. Products or facilities to be used by the elderly and/or the handicapped also deserve special attention.

In summary, many accidents are blamed, at least in part, on operator error. In many cases, the errors made by the operators were the result of poor human factors engineering in design. The designer set up the error-prone situation, and the operator was simply the victim, performing very consistently with the way any other human would have operated under the same conditions. This is not to say that operators do not make accident-causing errors or that all errors made by operators were the result of poor engineering or design. Some common types of operator errors are

Omit required actions
Perform nonrequired actions
Fail to recognize needed action
Respond improperly (early, late, wrong)
Poor communication
Maintenance error

Some common types of designer errors that contribute to operator errors are that they require operators to

Perform repetitious acts
Quickly recognize and correct problems
Perform rapid, complex computations
Operate in poor environment
Use improper tools
Pay continuous attention for long periods
Detect sensors outside human range
Perform physical tasks and talk simultaneously
Violate stereotypes
Perform correctly in life-threatening situations.

REVIEW QUESTIONS

1. Explain why human reliability is generally much lower and much more difficult to predict than hardware reliability.
2. Name the two principal types of stress that affect human reliability.
3. Name five failure modes that adversely affect human reliability.

4. List the steps in the system safety precedence sequence.
5. Give two examples of natural behavioral stereotypes and three examples of learned stereotypes.
6. What are the four areas to be considered in reviewing human factors in design?
7. Name three types of human errors made by operators.
8. Name five human factors errors made by designers.

REFERENCES

Johnson, William G. 1973. *MORT, the Management Oversight and Risk Tree.* Washington, DC: U.S. Atomic Energy Commission.

Johnson, William G. 1980. *MORT Safety Assurance Systems.* New York: Marcel Dekker.

McCormick, E. J. 1976. *Human Factors in Engineering and Design.* New York: McGraw Hill Book Co.

McElroy, F. F., ed. 1981. *Accident Prevention Manual for Industrial Operations.* 2 vol. Chicago: National Safety Council.

PART IV

SYSTEM SAFETY ANALYSIS TECHNIQUES

CHAPTER 13

Energy Trace and Barrier Analysis

Energy trace and barrier analysis (ETBA) is a relatively new technique based on some of the principles found in the management oversight and risk tree (MORT) program, which defines an *incident* as an unwanted flow of energy. An *accident* is defined as occurring when this unwanted flow of energy, in the absence of adequate barriers, strikes targets in the energy path and injures people and/or damages property (Fig. 13-1). So an incident is an unwanted energy flow and an accident is an unwanted energy flow or incident that results in adverse consequences.

The sides of the accident triangle that are discussed in the MORT program are the unwanted energy flow, barriers that are less than adequate to prevent or control the energy flow, and targets (persons or objects) in the energy path (Fig. 13-1). This approach is the basis for ETBA, which systematically analyzes these three factors and their interrelations.

PURPOSE OF ETBA

This particular analytical technique can be used to aid in preparing preliminary hazard lists (PHLs), conducting preliminary hazard analyses (PHAs), subsystem hazard analyses (SSHAs), or system hazard analyses (SHAs). The ETBA may be helpful in performing operating hazard analyses (OHAs) and accident analyses and in other situations. The ETBA seems to be particularly well-suited to facility system safety programs.

INPUT REQUIREMENTS

The primary input requirements for performing an ETBA are project plans and drawings showing the general facility (or project) layout. The amount of detail required depends on the type of analysis being performed.

System Safety for the 21st Century: The Updated and Revised Edition of System Safety 2000,
by Richard A. Stephans
ISBN 0-471-44454-5   Copyright © 2004 John Wiley & Sons, Inc.

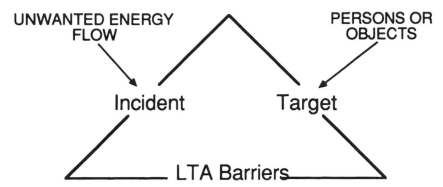

Figure 13-1   Accident triangle. (Source: MORT Workshop. DOE/SSDC.)

Preliminary drawings or sketches may be adequate to prepare a preliminary hazard list. More detailed drawings are required for a preliminary hazard analysis, and even more detail is required for subsystem and system hazard analyses. Analytical trees, copies of maintenance and operating procedures (if available), and site maps may also be helpful.

GENERAL APPROACH

The first step in conducting an ETBA is to identify the types of energy associated with the project.

Then, for each energy type, locate the point(s) at which the energy enters or originates, and trace the energy flow or energy path throughout the project.

Identify and evaluate the barriers that are in place to control the energy and the potential targets of any unwanted energy flow that could occur if the barriers fail.

Determine the risk associated with each potential unwanted energy flow and express this risk in terms of a risk assessment code (RAC).

Finally, recommend controls for unacceptable risks and for improving the overall safety of the project, and, if required, recommend systems or subsystems for further analysis.

INSTRUCTIONS

Collect the necessary input documents and reference resources, including applicable codes, standards, and regulations, list of consultants, lessons learned information, examples of ETBAs on similar projects, other analyses and/or a PHL on this project, and other materials that may aid in the ETBA effort).

List the types of energy associated with this project, which may include

Acoustical radiation

Corrosive

Electrical

Electromagnetic and particulate radiation

Explosive pyrophoric

Flammable materials

Kinetic-linear

Kinetic-rotational

Mass, gravity, height

Nuclear

Pressure-volume/K-constant-distance

Thermal (except radiant)

Thermal radiation

Toxic pathogenic

Complete the ETBA worksheets (Fig. 13-2) for each energy type. Use the completed ETBA worksheets to complete the appropriate type of analysis (PHL, PHA, SSHA, SHA) worksheets and report (see Chapter 8).

ENERGY TRACE AND BARRIER ANALYSIS

Project _____   Date _____

Prepared by _____   Page ___ of ___

Energy type_____

Drawing number (s)_____

ENERGY QUANTITY & LOCATION	BARRIERS	TARGETS	RAC	COMMENTS/BARRIER EVALUATION	RECOMMENDED ACTIONS	CONTROLLED RAC	STANDARDS

Figure 13-2   Energy trace and barrier analysis worksheet.

Converting ETBA worksheets into PHA (or other analysis) worksheets may or may not be necessary or desirable. Depending on the complexity of the project and the type of analysis performed, writing the analysis report directly from the information provided by the ETBA worksheets and including the ETBA worksheets as part of the report may be appropriate. Generating PHA (or other analysis) worksheets from the ETBA worksheets is relatively easy to accomplish because the "Energy Quantity & Location" column of the ETBA can be reworded to describe the "Hazardous Event" Column of the PHA worksheet (Fig. 8-2); the "Barriers" column of the ETBA worksheet can be reworded to correspond to the "Causal Factors" column; and the "Targets" column can be transformed into the "System Effects" column. Hazardous events are usually energy related; causal factors involve barrier failures; and systems effects are determined by the target of the unwanted energy flow.

Worksheet

Heading. Enter the project title and/or number, analyst's name, the type of energy being traced on this worksheet, and the appropriate drawing identification in the spaces provided. The drawing(s) are a part of the report or readily available to those using the analysis.

Column 1—Energy Quantity and Location. Describe the nature of the energy. Quantify it if possible by giving voltages, amps, watts, pressures, flow rates, container sizes, speeds, or weights. The location may be identified by narrative description (southeast corner of room 137; top right corner of drawing 222; output side of substation 2) and/or by coding (A1, A2, A3) and then cross-referencing to corresponding locations on drawings.

Column 2—Barriers. Describe all physical and procedural (if known) barriers in place at the indicated location to control or restrict any unwanted energy flows. For example, electrical barriers could include wire insulation, conduit, walls, fences, air space, and metal enclosures. Ground fault circuit interrupters (GFCIs), circuit breakers, and lightning arrestors are also electrical barriers that protect against overcurrent and overvoltage conditions. Additional electrical barriers could be warning signs and lockout or tagout procedures.

Column 3—Targets. List the persons or objects at this particular location that could be in the path of an unwanted energy flow. Quantify if possible (number of people, value of property). Include other systems and subsystems that could be adversely affected by the unwanted energy flow.

Column 4—RAC. Enter the risk assessment code associated with this particular unwanted energy flow. List a separate RAC for each potential target.

Column 5—Comments/Barrier Evaluation. Provide comments on the adequacy of the existing barriers to control potential unwanted energy flows at that particular location. If appropriate, include comments on compliance with applicable codes, standards, and regulations.

Column 6—Recommended Actions. Recommend steps to be taken to improve the safety of the project at this particular location. Recommendations could include rerouting the energy, reducing the energy level, improving barriers, adding additional barriers, moving or rerouting targets, and hardening targets. Recommendations are required for risk assessment codes that do not meet risk acceptability criteria (see Chapter 11).

Column 7—Controlled RAC. Enter the risk assessment code after the actions recommended in column 6 have been taken.

Column 8—Standards. Use this column to list any applicable codes, standards, or regulations.

REVIEW QUESTIONS

1. What is an *incident* (as defined in this chapter)?
2. What is an *accident*?
3. What are the sides of the accident triangle?
4. For what types of analysis is ETBA well suited?
5. What are the input requirements for an ETBA?
6. What general steps are involved in performing an ETBA?

REFERENCES

Hammer, Willie. 1972. *Handbook of System and Product Safety.* Englewood Cliffs, NS: Prentice-Hall.
Johnson, William G. 1973. *MORT, the Management Oversight and Risk Tree.* Washington, DC: U.S. Atomic Energy Commission.
Johnson, William G. 1980. *MORT Safety Assurance Systems.* New York: Marcel Dekker.
U.S. Army. (n.d.) Facility System Safety Manual Washington, DC: HQ Safety, USACE. Draft.

CHAPTER 14

Failure Mode and Effects Analysis

Failure mode and effects analysis (FMEA) is one of the better-known and more widely used systems safety techniques. The FMEA is used in Department of Defense, NASA, Department of Energy, and private industry programs. Even though different worksheets are used and it may also be called "Failure Mode and Effect Analysis," "Failure Modes and Effect Analysis," or "Failure Modes and Effects Analysis," the general approach to conducting an FMEA is relatively consistent. Additionally, depending on the organization, reference, or analyst, significant distinctions are sometimes made between an FMEA and a failure mode(s) and effect(s) criticality analysis (FMECA). In some organizations (NASA, for example), a critical items list (CIL) is developed from the FMEA.

In larger organizations where safety, reliability, maintainability, operability, and other factors may be considered separately by specific organizational units, the FMEA is widely recognized as a reliability tool, and rightfully so. The FMEA, by its very nature, specifically determines what can go wrong with each individual piece of hardware and what effects each failure can have. Technically, distinctions are not hard to make among those failures that affect safety, those that affect maintainability, and those that simply affect reliability. As a practical matter, however, most of the time something that breaks or fails to work properly has at least an indirect safety implication. In any event, recognizing that many organizations do not have separate reliability organizations, and that those that do generally have an established protocol for extracting safety-related data from the FMEA and using it in the system safety effort, it is included here as simply a system safety tool without concern for organizational turf or distinctions between FMEAs and FMECAs.

The FMEA evaluates reliability and identifies single-point failures. It can be performed at different levels and thus at different times in the life cycle.

The FMEA can be used as a top-level evaluation tool relatively early in the program by examining the ways in which subsystems can fail and the effect of

System Safety for the 21ˢᵗ Century: The Updated and Revised Edition of System Safety 2000, by Richard A. Stephans
ISBN 0-471-44454-5   Copyright © 2004 John Wiley & Sons, Inc.

those failures on the system or other subsystems. This use is commonly known as a *functional* FMEA.

The probably more common and more useful type is the *hardware* FMEA, which requires detailed design information and concentrates on assemblies, subassemblies, and components.

The FMEA evaluates the reliability of the design and identifies single-point failures that can lead to system failure. It also determines the possible effects of all potential failures and thus can aid in identifying failures with safety significance. It provides a cause-effect relationship for failures, some of which may be safety related. Failure rates or other reliability statistics can be used to quantify the FMEA.

The primary advantages of an FMEA are that critical single-point failures can be identified and that reliability can be evaluated in detail. It may identify areas or parts with poor reliability and allow early and cost-effective design changes. If it is performed functionally early in the project, it may reduce the amount of more detailed FMEA needed.

FMEA limitations include a failure to address the operational interface, multiple failures, or human factors.

PURPOSE OF FMEA

The FMEA is used primarily for doing subsystem and system hazard analyses, but a functional FMEA can also be useful in performing preliminary hazard analyses and operating hazard analyses. It helps to identify critical items in terms of safety and reliability.

INPUT REQUIREMENTS

Detailed design information is required to perform a hardware FMEA. Analytical trees, design drawings, and/or functional block diagrams are required so that the exact composition and/or configuration of each system, subsystem, assembly, and subassembly to be analyzed is clearly defined. As noted in Chapter 10, a well-constructed analytical tree is an excellent feeder document for an FMEA in that the configuration of the analytical tree matches the configuration of the project and the event symbols from the tree provide direct input for completing the "Component Description" column of the FMEA worksheets.

The functional FMEA requires less detail and can be performed with the upper branches of an analytical tree. The requirement is to identify subsystems and their functions, not details of subsystem configuration or design.

Failure rates or similar reliability statistics are required to quantify the analysis. This information may be difficult to obtain and even more difficult to validate.

GENERAL APPROACH

Failure mode and effects analysis, as the name implies, basically takes a system or subsystem, breaks it down into individual assemblies, subassemblies, and components, and then systematically looks at the different ways each component could fail and the effects of each failure on both the immediate assembly and eventually the entire system or subsystem. If possible, the effort includes quantified reliability data.

INSTRUCTIONS

First, collect the input documents and reference resources. The primary input documents include an analytical tree and/or block diagram, drawings, and narrative descriptions of the system. The most important resources are a data base, lessons learned file, historical data, and/or other source(s) of failure rates or other reliability statistics. The preliminary hazard list (PHL), preliminary hazard analysis (PHA) and any other hazard analyses should also be available.

The next steps in performing an FMEA are to define the system to be analyzed and to determine the scope of the analysis.

The scope is easily defined if an analytical tree is used as a feeder document (Chapter 10). The indenture levels can be identified by specifying the number of tiers of the tree to be included in the analysis (limits of resolution). A functional FMEA includes upper and/or middle tiers only; a hardware FMEA includes the entire tree.

The functional FMEA tends to use deductive logic, that is, to ask what the cause may be and to focus on the functional failure modes and their causes. It can be used early in the program and is usually done at the subsystem or assembly level. The hardware FMEA tends to be more inductive, asking "what if" and "when" questions, in that common component failure modes are listed and then the focus of the analysis is on the effects of each failure mode. It needs fixed design data and goes to the component level.

If the limits of resolution of the tree are broader than those desired for the analysis (that is, the system is being defined more narrowly), "trimming" some of the branches from the tree may be necessary.

If an analytical tree is not used, in addition to defining the system and clarifying the scope in terms of breadth (limits of resolution) and depth (indenture levels), a block diagram or other project description document may be required.

From the analytical tree (and/or other input documents), prepare a systematic breakdown of the system to the specified level of detail (generally subsystem or assembly level for a functional FMEA and component level for a hardware FMEA).

Complete the appropriate FMEA worksheets for the project (Fig. 14-1),

FAILURE MODE AND EFFECTS ANALYSIS

Page _____ of _____ Pages

System _____

Date _____

Subsystem _____

Analyst _____

Component Description	Failure Mode	Effects on Other Components	Effects on System	RAC or Hazard Category	Failure Frequency Effects Probability	Remarks

Figure 14-1   Failure mode and effects analysis worksheet.

and finally extract and use the information generated, which could include compiling a critical items list (CIL), writing a stand-alone hazard analysis report, and/or preparing a subsystem or system hazard analysis report by using relevant information from the FMEA.

Worksheet

Heading. Enter the name of the system (project) and the subsystem being evaluated on this page. Enter also the date the worksheet is completed and the name of the analyst completing the worksheet.

Column 1—Component Description. Enter the component description (narrative description and part number, if applicable). Note that this particular form was designed for a hardware FMEA. If a functional FMEA is being performed with this form, the "component" may be a subsystem, assembly, or subassembly. For some projects, breaking this column into two or three separate columns may be desirable. For example, if specific part numbers are assigned to all or most components and/or if separate item numbers are desired, a separate column (usually added as the first column with the component description becoming the second column) may be added just for item or part numbers. If the function of the component is not apparent or cannot be easily included in the narrative description, adding a separate "Function" column after the "Component Description" column may be necessary.

Column 2—Failure Mode. In this column, list all the ways in which the item in column 1 could fail to perform as designed. Common component failure modes Include failure to operate, operation at the wrong time (too early or too late), failure to stop operating, and partial operation.

Column 3—Effects on Other Components. List the impact of each failure listed in column 2 on other components. As in column 1, "components" may also include subassemblies, assemblies, or subsystems, especially if a functional FMEA is being performed. List those components that will physically be damaged and those whose functions may be adversely affected, which are generally those located adjacent to or near the failed component physically and/or operationally. They are generally in the same or the next higher indenture level or tier.

Column 4—Effects on System. For each failure mode listed in column 2, list the worst-case effect of that failure on the system. Responses in this column can range from "none" to "system failure."

Column 5—RAC or Hazard Category. Enter the risk assessment code (RAC) or hazard category for each failure listed in column 2. An RAC is recommended; however, some organizations evaluate FMEA failure only in terms of severity or criticality. In any event, the risk assessment codes or hazard categories must be clearly defined, and their definitions must be included in the FMEA report and/or physically attached to the FMEA worksheets.

Column 6—Failure Frequency Effects Probability. This column is optional. The failure frequency effects probability is a measure of the likelihood that the failure mode listed in column 2 will result in the system effect listed in column 4. This information is normally available only from lessons learned or other historical files and usually is not available for research and development projects. For many efforts, simple failure rates or mean times between failure data are substituted.

Column 7—Remarks. Use this column to add any information that may aid in evaluating the risk associated with the component being evaluated or any suggestion for improving the reliability or safety of the system. Comments on the source and reliability of data may also be appropriate in this column.

Appendix: Sample FMEA

The following is an abbreviated example of a failure mode and effects analysis on a simple wet pipe sprinkler system. Even though reliability data may be relatively accurate, it is provided for illustrative purposes only and should not be used for other purposes.

I. SUMMARY

This failure mode and effects analysis (FMEA) of the wet pipe sprinkler system for the _____ building at _____, _____, indicates that the basic system and related hardware are very reliable if installed and maintained properly. No further formal analysis of the system is recommended. However, it is important that applicable codes, standards, and regulations be closely followed in all phases of design, procurement, installation, testing, operations, and maintenance.

Particular care should be taken in specifying the type of thread connections for the fire department connection to ensure that the connection is compatible with local fire hoses.

A review of maintenance and operating instructions and procedures should be conducted with emphasis on ensuring that key components are properly maintained and tested.

It is extremely important that stringent control measures be taken to ensure that the main water control valve remain open because the primary cause of system failure (37%) is a closed water valve. Sprinkler head care is also important because virtually all sprinkler head failures are caused by painting, damaging, or obstructing the heads.

Although outside the scope of this analysis, a thorough study of the available water supply is also recommended. Another leading cause of system failure is inadequate water supply. Also, the relative importance of the fire department connection subsystem is determined largely by the adequacy of the main water supply. If the main water supply is marginal or unreliable, the fire department connection may be critical.

II. PROJECT DESCRIPTION

The system being examined in the FMEA is the wet pipe sprinkler system for the _____ building at _____, _____. (Insert brief building description here; include type, purpose, size, occupancy, and other relevant data, with one or two drawings.)

For more detailed information, see the project description brochure (PDB) and the preliminary hazard analysis (PHA).

The wet pipe sprinkler system consists of a single riser with water flow indicator and electric alarm feeding _____, _____ sprinkler heads (Fig. 14-2).

Water requirements are for a residual pressure of _____ psi, with a flow rate of _____ gpm at the base of the riser for _____ minutes.

Key Components

Valves

1. Control valve—wall post indication valve (PIV) or open screw & yoke (OS&Y) if control valve is not provided in underground system for each inside system.
2. Check valve if water flow alarm is used and no other check has been provided in the system.

Fire Department Connection—$2\frac{1}{2} \times 2\frac{1}{2} \times 4$ hose connection and check valve with ball drip.

Water Flow Indicator—A water flow indicator and electric alarm may be installed as a primary alarm in place of an alarm valve and water motor alarm; however, a check valve must be provided in the system.

System Piping—A system of piping progressively increasing in size in proportion to the number of sprinklers from the most remote sprinklers to the source of supply. The pipe size and distribution are determined from standards or hydraulic calculations as outlined by the NFPA, F.M. insurance underwriter, or other authorities for the hazard being protected.

Sprinklers—For types, orifice sizes, and temperature ratings, see the sprinkler head section of the data book. Sprinklers are spaced to cover a design-required floor area (from 36 to 200 square feet) as designated by the authorities indicated in system piping.

Drain—A flow test drain and main drain must be provided.

Operation

When a fire occurs, the heat produced will fuse a sprinkler, causing water to flow. The waterflow indicator is activated by the water flow. The paddle, which

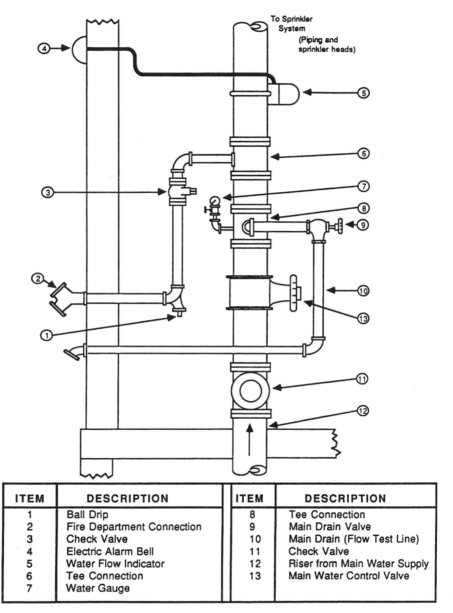

ITEM	DESCRIPTION	ITEM	DESCRIPTION
1	Ball Drip	8	Tee Connection
2	Fire Department Connection	9	Main Drain Valve
3	Check Valve	10	Main Drain (Flow Test Line)
4	Electric Alarm Bell	11	Check Valve
5	Water Flow Indicator	12	Riser from Main Water Supply
6	Tee Connection	13	Main Water Control Valve
7	Water Gauge		

Figure 14-2 Wet pipe sprinkler system: typical riser with water indicator.

normally lies motionless inside the pipe, is forced up, thereby activating the pneumatic time delay mechanism, which closes or opens a microswitch after the preset retard time has elapsed. This action causes an electric alarm to sound. The alarm will continue to sound as long as there is a flow of water in the system. The water will flow until it is shut off manually.

III. METHODOLOGY

The analytical technique used for this study was the failure mode and effects analysis (Figure 14-3 shows the FMEA worksheets filled out for the above "Sprinkler" example).

The FMEA is a procedure to examine a system systematically by breaking it into subsystems, components, and sometimes even individual parts or pieces to determine the potential failure mode of each item examined and the effects that each failure would then have on the system.

Used originally as a reliability tool, the FMEA is now often used to identify and prioritize safety problems associated with hardware failures. This is usually done by including a risk assessment code (RAC) in the analysis (Table 14-1). (Note: When a RAC or other method of quantifying is used to identify critical safety items, some organizations and analysts call the technique failure mode and effects criticality analysis [FMECA].)

HAZARD SEVERITY

I	Catastrophic	May cause death or loss of a facility
II	Critical	May cause severe injury, severe occupational illness, or major property damage
III	Marginal	May cause minor injury or minor occupational illness resulting in lost workday(s) or minor property damage
IV	Negligible	Probably would not affect personnel safety or health and thus less than a lost workday but nevertheless is in violation of specific criteria

MISHAP PROBABILITY

Frequent	A	Likely to occur frequently
Probable	B	Will occur several times in life of a facility

TABLE 14-1. Risk Assessment Code

	Severity			
Probability	I Catastrophic	II Critical	III Marginal	IV. Negligible
A Frequent	1	1	1	3
B Probable	1	1	2	3
C Occasional	1	2	3	4
D Remote	2	2	3	4
E Improbable	3	3	3	4

FAILURE MODE AND EFFECTS ANALYSIS

Page __1__ of __4__ Pages
Date __6 July 89__
Analyst __M. Wilson__

System __WET PIPE SPRINKLER__
Subsystem __MAIN WATER SUPPLY__

Component Description	Failure Mode	Effects on Other Components	Effects on System	RAC or Hazard Category	Failure Frequency Effects Probability	Remarks
Check Valve	Fails closed		System fails	2	<.001	
	Fails open		May contaminate water supply	3	<.001	
Water Control Valve	Fails open	Unable to test, repair	Unable to test, repair	3	<.001	
	Fails closed		System fails	2	=.001	
	Closed valve		System fails	1	.37	Greatest cause of failures
Sprinkler Heads	Fails to open	--	System fails (in local area)	2	.04	System failure if multiple failures (common cause)
	Opens prematurely	--	Water damage, system temporarily out of service	3	<.001	
	Inadequate spray pattern		Partial failure locally	3	<.001	No data (estimate)
Tee Connection/ Pipes	Rupture		Water damage, system out of service	2	<.001	Most likely on private, industrial water supplies. Readily detected.
	Leak		Reduced water pressure at sprinker heads	4	Unknown	
	Frozen/clogged		System failure	2	<.02	

Page __2__ of __4__ Pages
Date __6 July 89__
Analyst __M. Wilson__

System __WET PIPE SPRINKLER__
Subsystem __FIRE DEPT. CONNECTION__

Component Description	Failure Mode	Effects on Other Components	Effects on System	RAC or Hazard Category	Failure Frequency Effects Probability	Remarks
Fire Department Connection	Non-compatible or stripped threads	Prevents connection of fire hose	Supplemental water supply not available	2	<.01	Check thread specification with fire department(s). Important only if supplemental water is required (10% of cases)
	Clogged, dirty	May clog valves, sprinkler heads	May prevent or restrict supple-mental water flow - could clog other parts of system	2	<.01	More likely to occur in public buildings (tampering)
Ball Drip	Fails open		Water leak if supplemental water in use	3	Unknown	Insignificant if supplemental water not used
	Fails closed	Pipe will not drain--subject to freezing, corrosion	May prevent use of supplemental system	3	Unknown	More important in cold climates.
Check Valve	Fails closed	--	Prevents supple-mental water flow	2	<.001	Important only if supple-mental water is required (10% of cases)
	Fails open	--	Allows back flow, reduces available pressure	2	<.001	Easily detected

f-9-36

Figure 14-3   Failure mode and effects analysis.

FAILURE MODE AND EFFECTS ANALYSIS

System __WET PIPE SPRINKLER__
Subsystem __ALARM__

Component Description	Failure Mode	Effects on Other Components	Effects on System	RAC or Hazard Category	Failure Frequency Effects Probability	Remarks
Water Flow Indicator	Paddles break or stick	May clog spinkler head	--	3	Unknown	
		Alarm may fail		3	Unknown	
	Overly sensitive	Unwanted alarm signal	False alarm	4	Unknown	Reportedly relatively common on some systems
Electric Alarm Bell	Fails to operate	--	May result in excess water damage or no response by fire department	3	Unknown	
	False alarm			4	Unknown	

System __WET PIPE SPRINKLER__
Subsystem __DRAIN__

Component Description	Failure Mode	Effects on Other Components	Effects on System	RAC or Hazard Category	Failure Frequency Effects Probability	Remarks
Water Gauge	Incorrect reading		Unreliable inspection, test results	4	Unknown	Important only if undetected and in combination with low pressure.
Main Drain Valve	Fails open		Reduces available pressure	2	<.001	Readily detected
	Fails closed		Unable to test, repair	4	Unknown	

Figure 14-3 *Continued*

Occasional   C   Likely but possible to occur in life of a facility
Improbable   E   So unlikely that it can be assumed occurrence may not
    be experienced

References for this study included in *Automatic Sprinkler Systems Handbook (3d edition)*, edited by John K. Bunchard and published by the National Fire Protection Association (NFPA) (1987) and *Automatic Sprinkler and Standpipe Systems,* by John C. Bryan, also published by NFPA (1976). Probability data was interpolated from NFPA and other studies as reported in these publications. (Note: As previously noted, probability/reliability data in this example are for illustrative purposes only and may not be accurate.)

The risk assessment code (RAC) matrix and the hazard severity and probability indices and definitions used in this study were taken from EMX-XXXX (the Facility System Safety Manual).

REVIEW QUESTIONS

1. What are the advantages of an FMEA?
2. What are the limitations?
3. For what types of analyses are FMEAs well suited?
4. What are the two types of FMEAs? Discuss.
5. What is a CIL? What is the relation between an FMEA and a CIL?
6. What are the input requirements for an FMEA?

REFERENCES

Clements, P. L. 1987. *Concepts in Risk Management.* Sverdrup Technology, Inc.

Dhillon, B. S., and Singh, C. 1981. *Engineering Reliability: New Techniques and Applications.* New York: John Wiley & Sons.

Goldwaite, William H., and others. 1985. *Guidelines for Hazard Evaluation Procedures.* New York: American Institute of Chemical Engineers.

Hammer, Willie. 1972, *Handbook of System and Product Safety.* Englewood Chiffs, NJ: Prentice-Hall.

Henley, E. J., and Kumamoto, H. 1981. *Reliability Engineering and Risk Assessment.* Englewood Cliffs, NJ: Prentice-Hall.

Henley, E. J., and Kumamoto, H. 1985. *Designing for Reliability and Safety Control.* Englewood Cliffs, NJ: Prentice-Hall.

Malasky, S. W. 1982. *System Safety: Technology and Application.* New York: Garland Press.

National Aeronautics and Space Administration. 1970. *System Safety.* NHB 1700.1 (V3). Washington, DC: Safety Office, NASA.

National Aeronautics and Space Administration. 1987. *Methodology for Conduct of NSTS Hazard Analyses.* NSTS 22254. Houston, Texas: NSTS Program Office, Johnson Space Center.

National Aeronautics and Space Administration. n.d. *Safety, Reliability, and Quality Assurance Phase I & II Training Manual.* Houston: Office of SR&QA, Johnson Space Center.

Rogers, W. P. 1971. *Introduction to System Safety Engineering.* New York: John Wiley & Sons.

Roland, H. E., and Moriarty, Brian. 1981. *System Safety Engineering and Management.* New York: John Wiley & Sons.

U.S. Air Force. 1987. *System Safety Handbook for the Acquisition Manager.* SDP 127-1. Los Angeles: HA Space Division/SE.

U.S. Army. n.d. *Facility System Safety Manual* Washington, DC: HQ Safety, USACE. Draft.

U.S. Department of Defense. 1984 (updated by Notice 1, 1987). System Safety Program Requirements. MIL-STD-882B Washington, DC: Department of Defense.

CHAPTER 15

Fault Tree Analysis

Fault tree analysis (FTA) was developed by Bell Telephone Laboratories for the U.S. Air Force in 1962 and has been used as one of the primary system safety techniques since the system safety effort began. The Boeing Company was also an early pioneer in the use of fault tree analysis.

Fault tree analysis is a very detailed analytical technique for determining the various ways in which a particular type of failure could occur. Fault tree analysis is based on the negative analytical trees discussed in Chapter 10 and uses the same event and gate symbols (Fig. 15-1).

Rectangles are used as general event symbols, circles are used to show base events, and diamonds are used to show undeveloped terminal events. *And* gates are used to indicate that, in order to get an output, all inputs below that gate are required, and *or* gates are used to indicate that, in order to get an output, any one or any combination of the inputs is required.

There are two basic approaches to fault tree analysis. The qualitative approach is used to determine, using deductive logic, the ways in which the undesired top event could occur. The quantitative approach adds reliability or probability of failure data.

Fault tree analysis is one of the most meaningful system safety techniques available for systematically reducing the probability of an undesired event. It can also be one of the more expensive techniques because it requires a skilled and knowledgeable analyst and a considerable amount of time, especially if the project is complex and a quantitative approach is required.

The primary advantages of FTA are that it does produce meaningful data to evaluate and improve the overall reliability of the system and that it evaluates the effectiveness of and need for redundancy.

Its limitations are that the undesired event evaluated must be foreseen and all significant contributors to failure must be anticipated, that the effort may be very time-consuming and expensive, and that its success depends upon the skill of the analyst. Human factors, because of the multiple failure modes and

System Safety for the 21st Century: The Updated and Revised Edition of System Safety 2000, by Richard A. Stephans
ISBN 0-471-44454-5   Copyright © 2004 John Wiley & Sons, Inc.

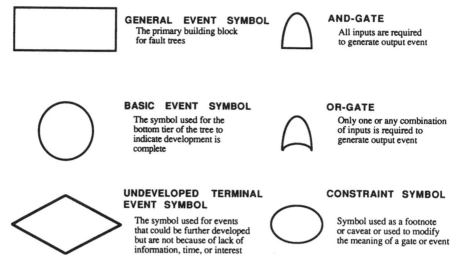

Figure 15-1 Fault tree symbols.

low levels of reliability discussed in Chapter 12, are normally not included (or at least not analyzed in depth) in FTA efforts.

PURPOSE OF FTA

The purpose of fault tree analysis is to determine, using deductive logic, the combinations of events that could cause a specified undesirable event to occur. This information is then used to evaluate the overall likelihood of occurrence and methods of reducing the likelihood of occurrence.

Fault tree analysis is used primarily as a tool for conducting system or subsystem hazard analyses, even though qualitative or top-level (that is, limited number of tiers or detail) analyses may be used in performing preliminary hazard analyses. Generally, FTA is used to analyze failure of critical items (as determined by a failure mode and effects analysis or other hazard analysis) and other undesirable events capable of producing catastrophic (or otherwise unacceptable) losses.

INPUT REQUIREMENTS

The major input requirement for a fault tree analysis is the top event. The nature of the undesired top event must be provided based on an earlier hazard analysis (formal or informal) and/or historical data.

Additionally, traditional project description documents (such as analytical trees, drawings, narrative descriptions, and block diagrams) are required to provide an in-depth knowledge of the project.

Prior hazard analyses (for example, preliminary hazard list, preliminary hazard analysis, and failure mode and effects analysis) should also be available.

For a qualitative FTA, reliability and failure probability data must be provided for all base events.

GENERAL APPROACH

The first step, of course, is to determine the nature of the failure to be analyzed and listed in the top block.

Once the particular type of failure is placed in that top block, then deductive logic and an in-depth knowledge of the system are used to determine the different ways in which that failure could occur. At each level, three particular types of component failures are examined. These are: primary failures, secondary failures, and command failures.

Examples of a primary failure (failure within design criteria):

- 1000-hour light bulb fails at 900 hours.
- Pressure vessel designed, tested, and certified for 500 psi ruptures at 380 psi.
- Helicopter rotor blades, with 1000 hour design life, develop cracks after 150 hours.

Examples of secondary failures (failure caused by factors external to the system):

- 1000-hour light bulb fails after 1100 hours (outside design life) or is broken by broom handle (broom not part of system) or fails due to overvoltage (exceeds design specifications; power source not part of system).
- Pressure vessel ruptured by forklift (external damage; forklift not part of system).
- Helicopter blades cracked by unauthorized flight maneuver (overload; pilot not part of system).

Examples of command failure (failure caused by different part of the system):

- Light bulb fails because of loose wiring (wiring part of system) or overvoltage (power source part of system).
- Pressure vessel ruptures because of overpressure from runaway pump (pump part of system).

- Helicopter blades cracked by unauthorized flight maneuver (pilot part of system).

One type is the primary failure, that is, the failure of that particular item, component, or piece of hardware within its design criteria. The second type of failure to be examined is secondary failure, which is caused by something external to the system. The third type of failure is command failure, which involves something internal to the system other than the failed component causing or being responsible for the failure.

After the fault tree is constructed with the symbols and construction steps listed in Chapter 10, the tree must be analyzed. If the tree is to be quantified, the appropriate reliability and failure probability data must be entered.

Cut sets and minimum cut sets should then be determined. A cut set is a group of base events, which, if they all occur, will cause the top event to occur. Using the rules of Boolean algebra, redundant base events can be eliminated and combined so that groups of base events containing the minimum number of events that could cause the top event to occur can be determined. They are called *minimum cut sets*. An examination of the minimum cut sets aids in identifying the base events that contribute most to the undesired top event and in determining the most effective ways to reduce the likelihood of top event occurrence.

INSTRUCTIONS

The first step in performing a fault tree analysis is to collect the appropriate project description documents, existing hazard analyses, and guidance documents and carefully review them to determine the limits, scope, and ground rules for the FTA. This review includes defining the system to be analyzed, the depth or indenture levels to be included in the effort, and, of course, the nature of the undesired event or failure to be studied.

The next step is carefully wording the undesired top event. This step is important because the wording of the top event determines, to a large extent, the scope of the effort. There are two basic approaches to limiting the scope and thereby focusing the effort. One approach is to state all of the conditions and restrictions in narrative form in the top block. The second approach is to use constraint symbols and/or undeveloped terminal events on the first tier.

For example, if the purpose of the FTA is to analyze the adequacy of design (from a safety viewpoint) of a nosewheel steering system on an aircraft and the undesired event of primary concern is a loss of directional control caused by primary failure of the nosewheel steering system when the aircraft is on takeoff or landing roll and traveling greater than tip-up speed (the speed at which an abrupt change in direction causes the aircraft to tip over physically, damaging wingtips and/or hardware suspended from wings) of 35 knots, the

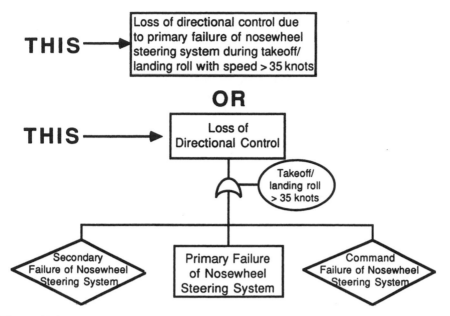

Figure 15-2 Top event examples. The wording and configuration of the top event determines the nature of the analysis and deserves careful attention. There are usually several acceptable ways to depict the same information.

top event block could read "Loss of directional control due to primary failure of nosewheel steering system during takeoff or landing roll with speed at or above 35 knots" (Fig. 15-2). An alternate way to provide the same focus would be to simply show "Loss of directional control" in the top block with "during takeoff/landing roll" and "speed >34 knots" shown as constraints, with "primary failure of nose-wheel steering" shown as a first-tier general event and "command failure (pilot error)," "secondary failure," and "other" shown as first-tier undeveloped terminal events.

Either approach or a combination is workable. Simply putting "loss of directional control" in the top block without further restrictions would produce a much more complex and costly effort than is required to meet the stated objective.

The next step is to construct the tree by using the construction steps outlined in Chapter 10 and including for each tier the command, primary, and secondary types of failure for each tier. The use of undeveloped terminal events depends upon the purpose of the analysis and the ground rules that have been established. For example, if the primary purpose of the effort is to evaluate the design of hardware only, focusing on primary hardware failures only may be appropriate, leaving command failures (those caused by something internal to the system other than the specific hardware being evaluated) and/or secondary failures (those caused by something external to the system) until later and thus showing them as undeveloped terminal events. This approach may be appropriate for a subsystem hazard analysis. The command and secondary failures may then be developed as part of the system (or integrated) hazard analysis. Note that whether a failure is "primary," "secondary," or "command" is determined largely by the way the "system" is defined.

Undeveloped terminal events should be used if the events do not represent significant risks (extremely reliable components), information is not available to develop the event, the event is not the responsibility of the analyst, or the event is not important at this time. Ideally, a complete, final analysis of a system does not contain undeveloped terminal events.

For example, if the simple lighting circuit shown in Figure 15-3 is self-contained, with power supplied by a 12-volt battery, the initial analysis of the circuit would probably concentrate on the overall configuration and primary failures of the power source (battery), fuse, switch, light bulbs, and wiring and connections. After ensuring that the reliability of the individual components and overall system meets overall requirements, examining the ways the system could fail due to damage from external sources (secondary failures) may be desirable. This analysis would probably require that the environment in which the system is to operate be defined. It then would concentrate on secondary failures such as exceeding the design environment of the components and the vulnerability of the bulbs to breakage. For example, expanding the branch of the tree (Fig. 15-4) containing only the light bulbs could expand the tree significantly (Fig. 15-4).

If all potential secondary and command failures for all components in this very simple system were analyzed in depth, even this effort could take on new dimensions. As a matter of fact, even the primary failure of a light bulb, shown in the example as a base event, could actually be expanded several more tiers if the light bulb were broken down into individual components. How tedious and time-consuming detailed fault tree analysis on complex systems can become can be readily appreciated.

In any event, after the tree has been developed consistent with the ground rules for the project, the next step is to analyze the tree to determine weak spots in the system, identify single-point failures, evaluate redundancy, and seek appropriate methods of improving the reliability and safety of the system.

If the tree is to be quantified, the appropriate reliability or probability of failure data are entered for the base events.

$$\text{Probability of failure } (P_F) = 1 - \text{Reliability } (R)$$

or

$$P_F = 1 - R$$

Analytical trees can be quantified using reliability (probability of success) or probability of failure data as long as numbers are consistent with the tree logic

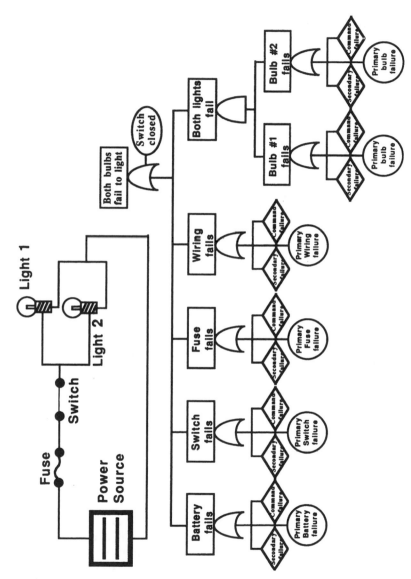

Figure 15-3   Example tree—primary events only. Early in the life cycle, the designer may concentrate on primary failures only.

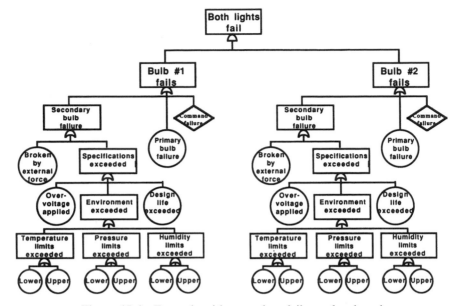

Figure 15-4 Example with secondary failures developed.

and are consistent with the ground rules for the analysis. Generally, reliability data are used with positive analytical trees, and failure probability data are used with negative or fault trees (Fig. 15-5).

For complex projects, determining cut sets and/or minimum cut sets is usually helpful in the analysis.

For quantified fault trees, the analysis obviously includes an analysis of the reliability and probability of failure data propagated from the base events to the top event.

As a practical matter, the modern-day analyst uses appropriate computer software to propagate quantified data and to determine cut sets and minimum cut sets. Computer assistance is also highly recommended for drawing and/or constructing fault trees.

For academic purposes, brief explanations of how to calculate probabilities and how to determine minimum cut sets follow. For the purist (or masochist) who chooses to attempt to perform fault tree analysis manually, a good statistics course emphasizing Boolean algebra is recommended.

Generally, probabilities under an *and* gate are multiplied, and probabilities under an *or* gate are added. As noted in Figure 15-6, adding probabilities under an *or* gate produces an approximation that generally is as accurate as the input data if, and only if, the probabilities being added are quite low, preferably below 0.1. If the failure probabilities are higher than 0.1, fault tree analysis should not be necessary. If the top event truly represents a loss of significantly high risk to warrant a fault tree analysis, then failure probabilities higher than 0.1 should be intuitively unacceptable.

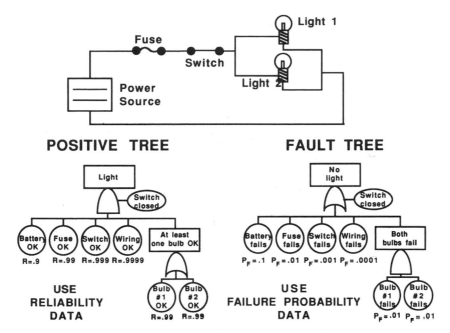

Figure 15-5   Positive *vs.* Fault Trees. For quantified efforts, the analyst frequently has the choice of using reliability data (numbers showing the probability that an item will work properly) or failure data.

MULTIPLY probabilities of input events under an "AND" gate to determine the probability of the output event.

$$P_A = P_B P_C P_D$$

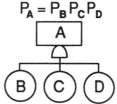

ADD probabilities of input events under an "OR" gate to determine the probability of the output event (Must also subtract probabilities of combinations - See below).

$$P_A = P_B + P_C + P_D - P_B P_C - P_C P_D - P_B P_D + P_B P_C P_D$$

IF "P's" are .1 or less, use $P_A = P_B + P_C + P_D$

NOTE: Rules apply to independent base events, that is the occurence of one will not effect the occurence of another

Figure 15-6   Probability calculation rules. These rules apply whether calculating the probability of success (reliability) or the probability of failure.

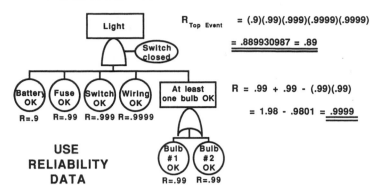

Figure 15-7 Positive tree with success probability calculated using reliability data.

Figure 15-7 illustrates how reliability (success probability) can be promulgated through a simple positive tree, and Figure 15-8 shows how failure probability (1 – reliability) can be promulgated through a fault tree. Note that the values are small enough to use approximations successfully in the fault tree example but would yield absurd results (probabilities greater than 1) in the positive tree example.

Cut sets and minimum cut sets can be determined manually by assigning letters to all gates in the tree and assigning numbers to all base events (circles). In both cases, start at the top of the tree and work down the tree from top to bottom. Then systematically build a matrix by replacing the letter for each gate with the combination of numbers and/or letters that are inputs for that gate.

For *and* gates, enter the input numbers and letters horizontally. For *or* gates, enter the input numbers and letters vertically in columns. For each number or letter added in a column, duplicate all the other entries from the parent row into the new row (Figs. 15-9 and 15-10). Repeat the process until the matrix contains numbers only.

Each row in the final matrix represents a cut set for the tree. If all events represented in any row were to occur, the top event would occur.

If certain base events appear more than once in the tree, the matrix can be simplified by eliminating all events that appear more than once in any given row and by eliminating any rows in which base events were contained in a shorter row. For example, if event 1 and event 2 were a complete row (cut set), other rows containing events 1 and 2 could be eliminated. The rows (cut sets) remaining after the matrix is simplified are called the *minimum* (or *minimal*)

FAULT TREE

NOTE: EXACT VALUE OF
$P_{F(Top\ Event)}$ = .110069013 = .11

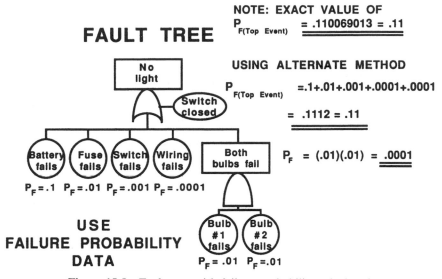

USING ALTERNATE METHOD

$P_{F(Top\ Event)}$ =.1+.01+.001+.0001+.0001

= .1112 = .11

P_F = (.01)(.01) = .0001

USE
FAILURE PROBABILITY
DATA

Figure 15-8   Fault tree with failure probability calculated.

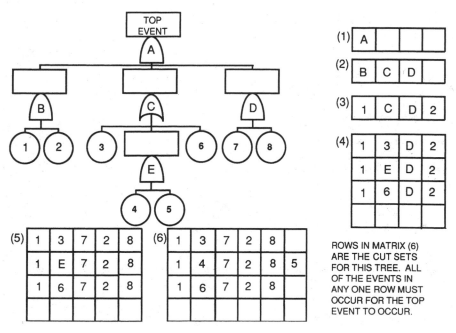

ROWS IN MATRIX (6)
ARE THE CUT SETS
FOR THIS TREE. ALL
OF THE EVENTS IN
ANY ONE ROW MUST
OCCUR FOR THE TOP
EVENT TO OCCUR.

Figure 15-9   Cut Set Example No. 1. In this example, each base event appears only once in the tree, therefore the cut sets derived are the minimum cut sets.

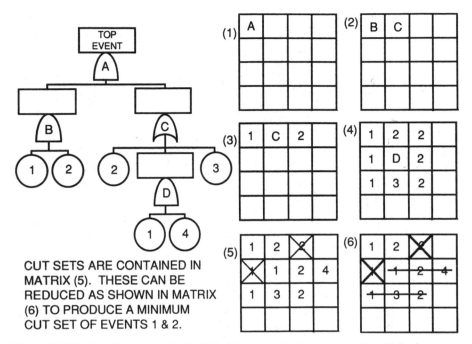

CUT SETS ARE CONTAINED IN
MATRIX (5). THESE CAN BE
REDUCED AS SHOWN IN MATRIX
(6) TO PRODUCE A MINIMUM
CUT SET OF EVENTS 1 & 2.

Figure 15-10   Cut Set Example No. 2. In this example, base events 1 and 2 both appear more than once, therefore it is possible to reduce the total cut sets by first eliminating redundant numbers in each row (matrix 5) and then eliminating rows which contain all of the events found in a shorter row (matrix 6) to produce the minimum cut set of events 1 and 2. This example uses the same fault tree configuration as one used by the Certified Safety Professional (CSP) examination in the *Examination Information* booklet, fifth edition, 1989 by the Board of Certified Safety Professionals (pp. 27–29).

cut sets and represent the minimum sets of events that, if they occurred, would generate the top event. The minimum cut sets can be used to draw a simplified fault tree. Cut sets and minimum cut sets can also be determined by using Boolean postulates to reduce the tree equations.

Appendix: Sample FTA

The following is an abbreviated example of a fault tree analysis on a simple wet pipe sprinkler system. Even though reliability data may be relatively accurate, it is provided for illustrative purposes only and should not be used for other purposes.

I. SUMMARY

This fault tree analysis (FTA) of the wet pipe sprinkler system for the _____ building at _____, _____, outlines the primary, secondary, and command failures of the system that would prevent it from extinguishing a fire in the facility.

The primary failures (failures that occur with design criteria) that could cause system failure include:

- Multiple failure of sprinkler heads in the fire area to open or multiple failure of sprinkler heads in the fire area to produce an adequate spray pattern.
- Failure of the water control valve or the main check valve in the closed position or water control valve indicating "open" when closed, causing improper positioning (PIV type valve).
- Failure of the fire connection check valve or the main drain valve, which allows unwanted water flow out of the system. This failure is critical only if the water supply is marginal and unable to overcome the loss.
- Primary failure of pipes and other components was not developed because such failure is highly unlikely and/or readily detectable.

Based on the high reliability of wet pipe sprinkler hardware, a primary failure of the system seems unlikely. Detailed analysis of primary failures is not recommended.

Most failures of sprinkler systems are the result of secondary failures; that is, the system fails because of influences external to the system.

Secondary failures that could cause system failure include:

- Exceeding the design environment for the system by subjecting an unprotected system to freezing temperatures, having a fire load (quantity or type of materials) that exceeds the design capacity of the system, or otherwise asking the system to do something it was not designed to do.
- Improper design of key components, of the system, or of the building.
- Improper installation of key components or of the system itself.
- Improper maintenance—failure to detect and/or properly correct problems or defects.

Even though the main water supply is not a part of the system being analyzed, obviously any failure of the main water system would cause system failure (and is a common cause of failure).

Other secondary failures, such as physical damage caused by heavy equipment, obstruction of sprinkler heads, or tampering, could also cause system failure.

Neither main water supply failure nor other secondary failures are fully developed because of the multiple failure modes possible and because analysis of these items goes beyond the scope of this analysis.

The final type of failure that could produce system failure is command failure; that is, a failure internal to the system "tells" the system not to work.

The most obvious and significant command failure for this system is the intentional closing of the water control valve. Less likely and readily detectable is the intentional opening of the main drain.

In summary, primary failures of wet pipe sprinkler systems are rare. Most systems fail because of secondary factors. The only significant command failure is the intentional shutoff of the main water supply.

No further formal analysis of the system is recommended. However, it is important that applicable codes, standards, and regulations be closely followed in all phases of design, procurement, installation, testing, operations, and maintenance.

A review of maintenance and operating instructions and procedures should be conducted with emphasis on ensuring that key components are properly maintained and tested.

It is extremely important that stringent control measures be taken to ensure that the main water control valve remain open because the leading cause of system failure (37%) is a closed water valve. Sprinkler head care is also important because virtually all sprinkler head failures are caused by painting, damaging, or obstructing the heads.

Although outside the scope of this analysis, a thorough study of the available water supply is also recommended.

II. PROJECT DESCRIPTION

The system being examined in this FTA is the wet pipe sprinkler system for the _____ building at _____, _____, _____. (Insert brief building description here; include type, purpose, size, occupancy, and other relevant data, with one or two drawings.)

For more detailed information, see the project description brochure (PDB) and the preliminary hazard analysis (PHA).

The wet pipe sprinkler system consists of a single riser with water flow indicator and electric alarm feeding _____, _____ sprinkler heads.

Water requirements are for a residual pressure of _____ psi, with a flow rate of _____ gpm at the base of the riser for _____ minutes.

Key Components

Valves

1. Control valve—wall post indication valve (PIV) or open screw & yoke (OS&Y) if control valve is not provided in underground system for each inside system.
2. Check valve if water flow alarm is used and no other check has been provided in the system.

Fire Department Connection—$2\frac{1}{2} \times 2\frac{1}{2} \times 4$ hose connection and check valve with ball drip.

Water Flow Indicator—A water flow indicator and electric alarm may be installed as a primary alarm in place of an alarm valve and water motor alarm; however, a check valve must be provided in the system.

System Piping—A system of piping progressively increasing in size in proportion to the number of sprinklers from the most remote sprinklers to the source of supply. The pipe size and distribution are determined from standards or hydraulic calculations as outlined by the NFPA, F.M. insurance underwriter, or other authorities for the hazard being protected.

Sprinklers—For types, orifice sizes, and temperature ratings, see the sprinkler head section of the data book. Sprinklers are spaced to cover a design-required floor area (from 36 to 200 square feet) as designated by the authorities indicated in system piping.

Drain—A flow test drain and main drain must be provided.

Operation

When a fire occurs, the heat produced will fuse a sprinkler, causing water to flow. The waterflow indicator is activated by the water flow. The paddle, which normally lies motionless inside the pipe, is forced up, thereby activating the pneumatic time delay mechanism, which closes or opens a microswitch after the preset retard time has elapsed. This action causes an electric alarm to sound. The alarm will continue to sound as long as there is a flow of water in the system. The water will flow until it is shut off manually.

III. METHODOLOGY

This system hazard analysis of a wet pipe sprinkler system was conducted using fault tree analysis (FTA).

The fault tree is a symbolic diagram showing the cause-effect relationship between a top undesired event and one or more contributing causes. It is a deductive analytical means to identify all failure modes contributing to the potential occurrence of a given top undesired event.

The "undesired top event" in this case was "failure of the system to extinguish fire using primary water supply."

Fault tree analysis is a technique by which the system safety engineer can rigorously evaluate specific hazardous events. It is a type of logic tree that is developed by deductive logic from a top undesired event to all subevents that must occur to cause it. It is primarily used as a qualitative technique for studying hazardous events in systems, subsystems, components, or operations involving command paths. It can also be used for quantitatively evaluating the probability of the top event and all subevent occurrences when sufficient and accurate data are available. Quantitative analyses shall be performed only when it is reasonably certain that the data for part/component failures and human errors for the operational environment exist.

In this study, as for most facility systems, qualitative analysis was appropriate.

In fault tree construction, specfic symbols are used to present events and logic gates. The use of symbology primarily assures consistency through the fault tree, aids in the identification and reference of events and logic gates, and simplifies the logic projected by the fault tree. Only the basic FTA symbol and tree development techniques are described here. The basic or primary symbols are as follows:

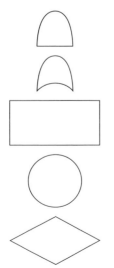

The *and* gate describes the logical operation whereby the coexistence of all input events is required to produce the output event.

The *or* gate defines a situation whereby the output event will exist if one or more of the input events exist.

The rectangle identifies an event that results from the combination of fault or failure events through an input logic gate.

The circle describes a primary failure event that requires no further development.

The diamond describes an event that is not further developed because of insufficient information or it is not of sufficient consequence.

Suitable mathematical expressions representing the fault tree entries may be developed using Boolean algebra. When more than one event on a chart can contribute to the same effect, the chart and the Boolean expression indicate whether the input events must all act in combination (*and* relationship) to produce the effect, or whether they may act singly (*or* relationship). The probability of failure of each component or of the occurrence of each condition or listed event is then determined. These probabilities may be from actual failure rates; vendors' test data; comparison with similar equipment, events, or conditions; or experimental data obtained specifically for the system. The probabilities are then entered into the simplified Boolean expressions. The probability of occurrence of the undesirable event being investigated may then be determined by calculation. When an FTA is used for qualitative analysis, care is required in the description of each event to be sure it can be fitted with a suitable probability.

The FTA analyzes three specific types of fault or failure: primary, secondary, and command.

A primary failure occurs when a component fails within its design criteria, that is, when it is being used properly but fails to do what it was designed to do. For example, if an incandescent light bulb designed to burn for 1,000 hours fails under normal conditions after only 900 hours, it is a primary failure.

A secondary failure occurs when something external to the system causes the component failure. For example, if our incandescent light bulb fails because it is broken by a broomstick, it is a secondary failure.

A command failure occurs when something internal to the system causes the component to fail. For example, if our incandescent light bulb fails to burn because the light switch is off, it is a command failure.

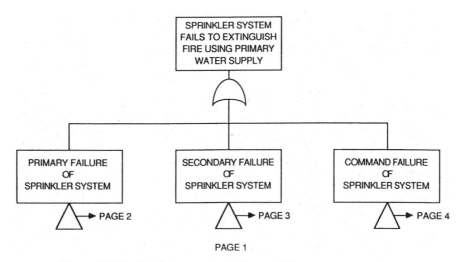

Figure 15-11   Fault tree analysis for sprinkler system example.

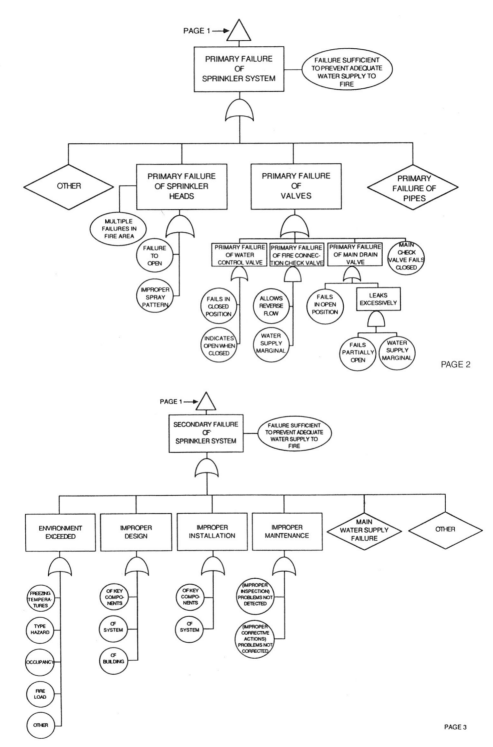

Figure 15-11   *Continued*

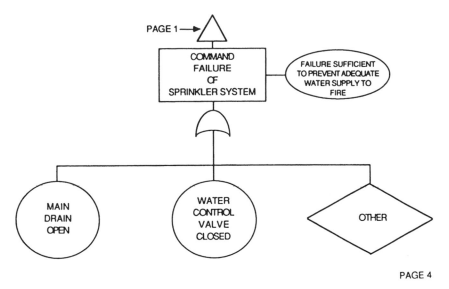

Figure 15-11 *Continued*

Fault tree analysis, expecially quantitative or detailed analysis, can be very tedious and time-consuming and thus very expensive. It should be reserved for extremely high hazard situations.

For most facility systems, if an FTA is appropriate, a top-level, qualitative effort is appropriate.

REVIEW QUESTIONS

1. What are the advantages of fault tree analysis?
2. What are the limitations of fault tree analysis?
3. At what point in the life cycle are FTAs generally done? For what types of hazard analysis are FTAs most appropriate?
4. What are the three types of component failure to be analyzed? Discuss and give examples of each.
5. What are the input requirements for fault tree analysis (qualitative and quantitative)?
6. What is the first step to be accomplished in performing an FTA?
7. Under what circumstances should undeveloped terminal events be included in an FTA?
8. Outline the manual method for calculating the probability that the top event will occur, given the failure probabilities of the base events (all <.02).
9. Outline the general manual method for determining cut sets.

10. Explain the general procedure for reducing a list of cut sets to a list of minimum cut sets.

REFERENCES

Barlow, R. E., Fussell, J. B., and Singpurwalla, N. D. *Reliability and Fault-Tree Analysis.* Society for Industrial and Applied Mathematics.

Buys, J. R. 1977. *Standardization Guide for Construction and Use of MORT-Type Analytical Trees.* ERDA 76-45/8; SSDC-8, Idaho Falls, ID: Energy Research and Development Administration.

Clements, P. L. 1987. *Concepts in Risk Management.* Sverdrup Technology, Inc.

Crosetti, P. A. 1982. *Reliability and Fault Tree Analysis Guide.* DOE 76-45/22; SSDC-22. Idaho Falls, ID: Department of Energy.

Hammer, Willie. 1972. *Handbook of System and Product Safety.* Englewood Cliffs, NJ: Prentice-Hall.

National Aeronautics and Space Administration. 1970, *System Safety.* NHB 1700.1 (V3). Washington, DC: Safety Office, NASA.

National Aeronautics and Space Administration. 1987. *Methodology for Conduct of NSTS Hazard Analyses.* NSTA 22254. Houston: NSTS Program Office, Johnson Space Center.

Roland, H. E., and Moriarty, Brian. 1981. *System Safety Engineering and Management.* New York: John Wiley & Sons.

U.S. Army. (n.d.) *Facility System Safety Manual* Washington, DC: HQ Safety, USACE. Draft.

U.S. Department of Defense. 1984 (updated by Notice 1, 1987). *System Safety Program Requirements.* MIL-STD-882B. Washington, DC: Department of Defense.

Vesely, W. E., and others. 1981. *Fault Tree Handbook* NUREG-0492. Washington, DC: U.S. Government Printing Office.

CHAPTER 16

Project Evaluation Tree

Project evaluation tree (PET) analysis is a very new technique. It is an adaptation of a program originally developed for the U.S. Air Force's Tactical Air Command (TAC) in 1988. The TAC's ground safety community had been involved with management oversight and risk tree (MORT)—based training for several years. A continuing point of concern was the complexity of the MORT chart's 1,500 different events and multiple transfers. Even though the value of the MORT chart as an accident investigation tool was evident, it seemed like overkill for use on many of the types of mishaps investigated by ground safety personnel. Thus, a tool was desired that would use a MORT-type approach but would be quicker and simpler to learn and use, could be tailored for use by single investigators (as opposed to boards of investigation), and would be written in air force terms.

The program developed was the Combat-Oriented Mishap Prevention Analysis System (COMPAS). It used two analytical charts. One, the COMPAS A chart, was basically a big positive tree showing the overall organization of a typical TAC wing, broken down organizationally and then broken down into the people, procedures, and facilities and hardware for that particular part of the organization.

The COMPAS B chart was an analytical tree that contained evaluation criteria to be applied to the personnel, procedures, and/or facilities and hardware identified or selected for evaluation on the A chart. It could be used as an inspection, accident investigation, or operational readiness tool.

In 1989, key concepts from TAC's COMPAS program were integrated in the U.S. Air Force's advanced occupational safety course, and the COMPAS B chart was revised for general use as the project evaluation tree (PET). Later that year, PET was introduced to NASA at the Johnson Space Center, to the general occupational safety community at the National Safety Congress in Chicago, to safety engineering students at the University of Houston (Clear

System Safety for the 21ˢᵗ Century: The Updated and Revised Edition of System Safety 2000, by Richard A. Stephans
ISBN 0-471-44454-5   Copyright © 2004 John Wiley & Sons, Inc.

Lake), and to U.S. Army Corps of Engineers safety professionals at their conference in Bethesda. In 1990, PET was included in courses presented to DOE-contracted personnel at the Nevada Test Site.

The project evaluation tree is an analytical tree to be used primarily as a graphic check in basically the same manner as the management oversight and risk tree (see Chapter 18). The PET chart, however, contains fewer than 200 event symbols and no transfers; the MORT chart contains approximately 1,500 events symbols, multiple transfers, transfers within transfers, and drafting breaks. The PET chart is divided into three branches: procedures, personnel, and plant and hardware.

It is self-tailoring in that for a simple inspection, mishap, or analysis involving only a few procedures, personnel, and/or facilities or items of hardware, the PET chart can be applied to the relevant items relatively quickly. For a complex situation involving many procedures, personnel, and/or facilities and items of hardware, many iterations of the PET chart are required, thus producing an in-depth analysis.

In side-by-side comparisons with the MORT chart and the mini-MORT chart (see Chapter 18), PET has consistently been more favorably received than either of the others by persons with and without prior MORT experience or training.

Except for course materials prepared by the author, this book is the only available reference for PET.

PURPOSE OF PET

The purpose of the project evaluation tree is to provide a relatively simple, straightforward, and efficient method of performing an in-depth evaluation or analysis of a project or operation. It is best suited for performing operating hazard analysis and accident analysis. It can also be a valuable review and inspection tool. If adequate information is available, PET analysis may be helpful in performing preliminary hazard analysis, subsystem hazard analysis, and system hazard analysis.

INPUT REQUIREMENTS

A PET analysis requires detailed information on the procedures, personnel, and facilities and hardware to be evaluated. In addition to actual procedures, upstream documents requiring the procedures and outlining the protocol for their generation, review, distribution, and updating are also required.

Job descriptions, organizational charts, training records, course curriculae, course materials, interviews, and other data are required for personnel evaluation.

Drawings, procurement documents, specifications, test plans and records, system safety plans and records, hazard analyses, and budget data may be required to evaluate facilities and hardware.

A positive analytical tree of the system or project to be evaluated is extremely useful, especially if organized into major branches containing procedures, personnel, and facilities and hardware (see Chapter 10).

The scope and depth of the analysis dictate the need for input data. The quality of the input data greatly influences the quality of the analysis.

GENERAL APPROACH

Basically, the PET chart is just a graphic checklist. The general procedure is to identify each procedure, individual and/or organization, facility, and piece of equipment or hardware to be analyzed and then systematically use the appropriate branch of the PET tree, in conjunction with the PET user's guide, to evaluate each part of the system or project (Fig. 16-1).

Color coding of the PET chart is recommended, using red for evaluation criteria determined to be less than adequate (LTA), green for items that are adequate, black for criteria or branches of the PET chart than do not apply to the particular project or item being evaluated, and blue to indicate areas with insufficient input data to make a decision. (This color-coding system is the same that is used for the MORT chart and the mini-MORT chart (see Chapter 18).

The PET chart is used primarily as a working tool to guide the analyst through the analysis. Information, especially items identified as less than adequate (red) is transferred to a PET analysis worksheet (see Fig. 16-6). These worksheets are then used like other hazard analysis worksheets for hazard tracking and preparing a narrative summary of the analysis.

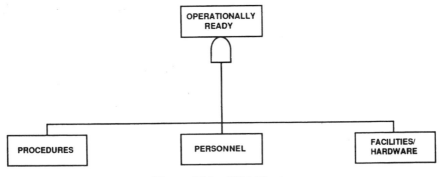

Figure 16-1 PET Chart.

The general approach for performing a PET analysis is basically the same whether performed as an operating hazard analysis, accident analysis, review, or inspection. The primary difference is in the nature of the input. For an operating hazard analysis, for example, the only procedures to be evaluated are generally the maintenance, inspection, and operations for the facility or hardware item being designed. The personnel evaluation may be limited to an examination of the organizational structure, selection requirements, and training requirements for personnel to occupy and/or service and operate the facility or equipment.

An accident investigation or analysis, however, also examines the specific procedures and the individuals involved in the accident.

Certain types of review or inspection may require only one branch of the PET tree; for example, the procedures branch can be used for performing procedural audits without using the rest of the tree.

INSTRUCTIONS

First, as with other analyses, define the scope of the analysis and the ground rules. Then collect the appropriate input data; organize the data into procedures, personnel, and facilities/hardware; and systematically evaluate the system or project using the PET user's guide and the PET chart. Note that multiple copies of the PET chart may be required.

Complete PET analysis worksheets for each item evaluated and use these worksheets to prepare the narrative summary of the analysis.

For accident analyses and other efforts for which input information may be provided piecemeal, start the PET analysis as early as practical and update it as new information is available. Use the PET chart as a guide for organizing the accident investigation and collecting evidence (Figs. 16-2 through 16-5).

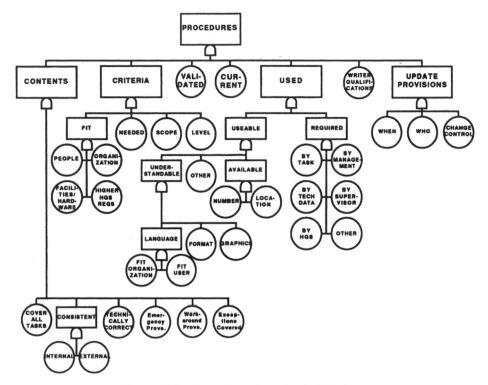

Figure 16-2   Procedures branch of PET chart.

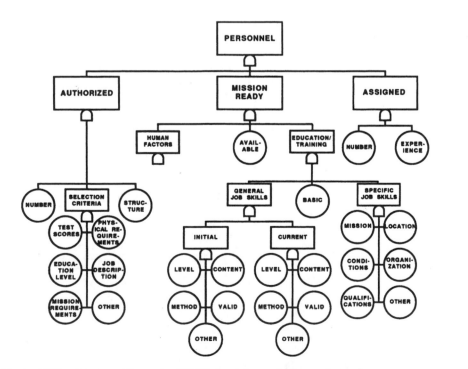

Figure 16-3   Personnel branch of PET chart (except human factors).

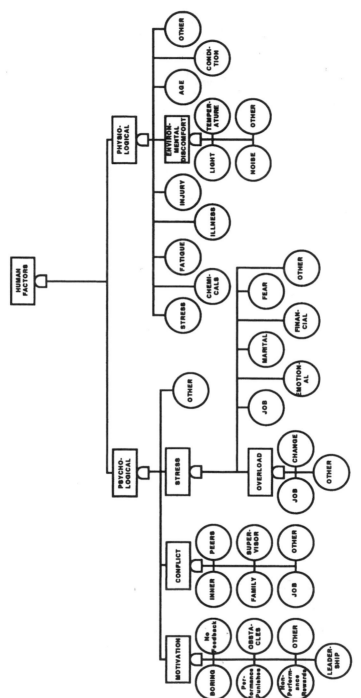

Figure 16-4   Human factors portion of personnel branch of PET chart.

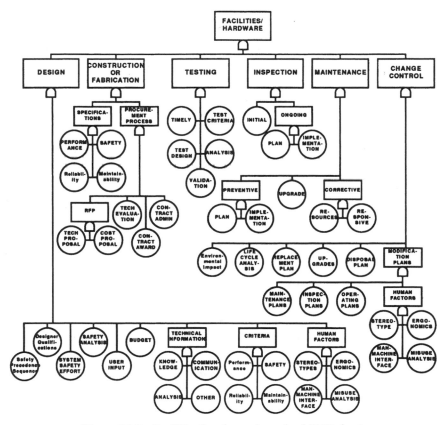

Figure 16-5   Facilities/hardware branch of PET chart.

PET ANALYSIS WORKSHEET

Prepared by:_____
Office Symbol:_____
Date: _____

Purpose:
Operational Readiness _____   Change Analysis _____   Inspection/Evaluation _____
Mission Failure _____   Accident Investigation _____

ITEM NO.	ITEM EVALUATED	PET EVENT	COLOR	PROBLEM/COMMENTS	RESPONSIBLE PERSON/AGENCY	STATUS	FINAL COMPLETION DATE

Figure 16-6 PET worksheet.

Appendix: PET User's Guide

The following information is provided to assist in understanding the nature of the questions to ask in evaluating the projects or programs using PET analysis.

This supplement is intended as a guide only. The analyst should not hesitate to expand or modify the data collection process to fit specific needs and circumstances.

Procedures

All codes, standards, regulations, operating procedures, maintenance procedures, and the like relevant to a given operation.

Contents

Do the procedures cover the subject?

- Cover all tasks. Are instructions available for all tasks needed to do the job correctly?
- Consistent. Does the document contradict itself or is it in conflict with other technical data?
  - Internal—Is the document free of confusing, different, or contradicting information or instruction? Are drawings, figures, and charts consistent with the text?
  - External—Are the requirements and directions on this document consistent with regulations, operating manuals, tech orders, checklist, and other relevant codes, standards, regulations, and technical data?
- Technically correct. Does the document explain what to do when things go wrong? Is the information factual and accurate? Does it really show the way things should be done?
- Emergency provisions. Does the document explain what to do when things go wrong? Are the emergency provisions adequately highlighted (bold print, separate page, special border)?

- Work-around Provisions. When certain requirements cannot be met or things cannot be done "normally" or in the desired manner, are alternate solutions or approaches addressed?
- Exceptions covered. Are there times, circumstances, or conditions when part or all of the document does not apply? Are the criteria for exceptions clearly stated?

Criteria

Do the procedures contain the necessary information in the correct language and format for successfully linking people, facilities, and hardware to provide a safe, efficient operation?

- Fit. Does the procedure accurately describe/outline the tasks as they are performed? Are the tasks arranged in a logical, efficient manner and described clearly?
- People. Were the procedures written for use by the people currently using them?
- Organization. Were the procedures written for the size and type of organization that is using them? Were they written for the level within the organization that is using them?
- Facilities/hardware. Were the procedures written for use with the facilities and hardware with which they are being used?
- Higher headquarters regulation. Are the procedures consistent with and in compliance with relevant regulations and requirements from higher levels (corporate office, government)?
- Needed. Is this piece of paper necessary? Does it provide information not available elsewhere that is needed to perform tasks?
- Scope. Is the procedure found in the appropriate level of the organization, written for this level, and locally supplemented as necessary?
- Level. Is this procedure found in the appropriate level of the organization, written for this level, and locally supplemented as necessary?

Validated

Have qualified people reviewed, checked, and tested the procedures to ensure that they are doing their job properly?

Current

Is there a simple method of ensuring that the procedure is current?

Used

Are the procedures being followed?

- Usable. Are the procedures so written and located that they are easy to understand and readily available?
  - Understandable
    - Language. Is the procedure written in the native language of the users with terms familiar to the organization and simple enough to be understood by the least-educated user?
    - Fit organization. Are acronyms, terms, and references specifically tailored to the using organization?
    - Fit user. Is the sentence structure and writing level simple enough for the least-educated user to understand what is being said? Are short words and short sentences used to explain technical or complex items?
    - Format. Is the size of the page and the print suitable for the conditions under which the procedure is to be used? Is the procedure well illustrated with pictures and diagrams to show what must be done? Can the procedure be easily updated? Is the procedure modular so that all users do not have to have the whole document?
    - Graphics. Are the graphics large enough and clear enough to be easily understood? Are they located with the supporting text?
    - Other. What other characteristics of the procedure make it difficult to use? Color? Shape? Type of print? Quality of paper? Fragile? Large? Heavy? Complex?
  - Available. Is the procedure located when and where it is needed with enough copies for all users?
    - Number. Are there enough copies for users, supervisors, and staff? Are additional copies readily obtainable to replace lost, worn, or damaged materials?
    - Location, Is the procedure available at the point of use? Is there a "place" for the procedure at its point of use?
- Required. Is the use of the procedure mandatory, and is the requirement enforced?
  - By tasks. Does the nature of the tasks indicate that a procedure should be used?
  - By management. Does management hold supervision accountable for the use of procedures?
  - By technical data. Is the use of the procedure a written requirement that is known to users?
  - By supervisor. Does supervision enforce the use of prescribed procedures?

- By headquarters. Are there specific requirements for use by a higher headquarters?
- Other. Are there state, local, or federal requirements for use of these procedures?

Writer Qualifications

Are the people producing the procedure qualified in terms of in-depth knowledge of the operation, knowledge of the intended use of the procedure, and knowledge of the users? Do they have adequate academic and technical education, experience, and writing skill? Do they have the authority to produce the document?

Update Provisions

Is there a workable system for revision of the procedure?

- When. Is it clear how often and under what circumstances the procedure must be reviewed and/or revised? Do the update provisions cover both "periodic" and "as needed" revision?
- Who. Who is responsible for determining when the procedure should be updated? Who is responsible for actually making revisions/updates? Who is responsible for revising the update process and outlining validation requirements?
- Change control. Are there provisions for automatically updating procedures when personnel or hardware changes occur? Are there provisions for automatically updating procedures when upstream requirements change? Are there provisions for making appropriate downstream changes? How are changes distributed and verified? Are provisions included for purging rescinded or outdated procedures from the system?

Personnel

Are the personnel required for mission accomplishment authorized, assigned, and mission-ready?

Authorized

- Number. Is the staffing level adequate for the mission to be performed under mission conditions?
- Selection criteria. Are the selection criteria matched to actual job requirements in such a way as to provide a minimum of performance discrepancies?

- Test scores. Are test scores consistent with the requirements for the task to be performed?
- Physical requirements. Are the physical requirements consistent with the physical requirements of the job?
- Education level. Are prerequisite education requirements consistent with actual task requirements?
- Job description. Are job descriptions consistent with actual work requirements?
- Mission requirements. Are specific job requirements consistent with the original selection criteria?
- Other. Are there other selection criteria that should be considered for this mission or task? If so, has it been done?
- Structure. Is the rank/grade structure adequate to ensure adequate experience, maturity, and depth of knowledge under mission conditions?

Mission Ready

Are the assigned personnel available, sufficiently trained, and in the proper mental, physical, and emotional condition to perform tasks correctly under existing conditions?

- Human factors. Are personnel mentally, physically, and emotionally fit to perform assigned tasks?
  - Psychological. Are personnel psychologically fit to perform job tasks?
    - Motivation. Do personnel really want to perform job tasks properly?
      - Boredom. Is the job repetitive, dull, and uninteresting? Does the job lack challenge to the point that success is meaningless?
      - No feedback. Is there any measure of performance? Do people know how they are doing? Are there any consequences?
      - Performance punishment. Is doing a good job rewarded by more and tougher assignments?
      - Obstacles. Are there personality conflicts, organizational red tape, conflicts of interest, or other obstacles to performance?
      - Nonperformance rewards. Are nonperformers expected and allowed to do less and work only on easy tasks?
      - Other. Are there other factors affecting motivation?
      - Leadership. Does the supervisor inspire subordinates by providing positive leadership and setting the example?
    - Conflict. Are job tasks not being performed properly because personnel are faced with some type of conflict?

- Inner. Are conflicting personal values, goals, or objectives degrading performance?
- Peers. Are disagreements or other conflicts with fellow workers degrading performance?
- Family. Are disagreements or other conflicts with family members degrading performance?
- Supervisor. Are disagreements or other conflicts with supervisors degrading performance?
- Job. Is a conflict with the goals, objectives, value, or methods of performance of the job itself degrading performance?
- Other. Is there another source of conflict degrading performance?
- Stress. Is each individual free from debilitating stress?
  - Overload. Are individuals overstressed as a result of having too much to do in the time allotted?
  - Job. Are job requirements unrealistic for existing conditions? Are requirements excessive in terms of quantity of work, quality of work, or both?
  - Change. Are rapidly and/or dramatically changing conditions and/or requirements overloading personnel?
  - Other. Are other factors placing overload-related stress on personnel?
- Job. Are individuals matched to jobs in such a way that they can deal with job stress and perform satisfactorily under existing conditions?
- Emotional. Are individuals free from debilitating emotional stress that can degrade performance, including love, infatuation, hate, and depression?
  - Marital. Are individuals free from debilitating marital stress? Marital stress can occur at any time but is usually most significant for the newly married, those separated from spouse (voluntarily or involuntarily), those suspecting infidelity, and those in the divorce process.
  - Financial. Are individuals free from significant stress resulting from excess financial obligations and/or insufficient resources to meet obligations?
  - Fear. Are individuals free from debilitating fear? Normally, significant stress from fear is associated with combat activities and/or other occasions where sudden and violent deaths or traumatic injuries may have occurred.
  - Other. Are other sources of personal stress significantly impairing job performance?
- Other. Are other psychological factors adversely affecting performance?

- Physiological. Are physiological factors preventing personnel from being mission-ready?
- Stress. Is some sort of physical stress such as positive or negative G-loading present to such a degree that task performance may be affected?
- Chemicals. Are drugs or alcohol being used/abused so that performance is adversely affected?
- Fatigue. Are individuals too tired to perform correctly?
- Illness. Are medical problems, diagnosed or not, adversely affecting performance?
- Injury. Are injuries preventing personnel from properly performing job tasks?
- Environmental discomfort. Is the work environment controlled adequately?
  - Light. Are lighting levels appropriate for the tasks being performed?
  - Temperature. Is the temperature within acceptable limits for adequate performance of job tasks.
  - Noise. Are noise levels controlled to prevent hearing damage and degradation of performance?
  - Age. Has age degraded the motor skills, reflexes, memory, sense, or other capabilities to the point that job tasks can no longer be performed properly?
  - Condition. Are personnel in adequate physical condition to perform assigned tasks under existing conditions?
  - Other. Are other physiological factors adversely affecting performance?
- Available. Are the assigned personnel available to perform mission tasks or are they on vacation, in school, incapacitated, on special assignment, or otherwise not present to perform assigned job tasks?
- Education and training. Have personnel been provided with the education and training needed to perform job tasks under existing conditions?
  - General job skills. Have personnel adequately mastered and maintained the general job skills necessary to perform job tasks under existing conditions?
    - Initial. Were job skills initially mastered?
      - Level. Was the initial general job training done in sufficient depth and at the appropriate skill level to prepare personnel to perform job tasks under mission conditions?
      - Contents. Did the initial general job training develop the skills required to perform job tasks under mission conditions?
      - Method. Were the methods of instruction used adequate for efficiently and effectively teaching required job skills?

- Valid. Were performance examinations conducted to ensure that required job skills were developed?
- Other. Are other factors affecting the adequacy of general job skill training?
- Current. Have job skills been adequately updated and upgraded?
  - Level. Has update/upgrade job training been in sufficient depth and at the appropriate skill levels to maintain job skills and knowledge necessary to perform job tasks under mission conditions?
  - Content. Have update/upgrade training courses and materials kept pace with changes in equipment, procedures, and requirements?
  - Method. Were appropriate methods of instruction used to provide update/upgrade job skills?
  - Valid. Were performance examinations conducted to ensure that training objectives were met?
  - Other. Are other factors involving update/upgrade training of general job skills hindering performance?
- Basic. Do personnel have the basic skills and knowledge necessary to function as members at the organization?
- Specific job skills. Have personnel been trained to perform specific tasks required under the specific conditions for specific missions?
  - Mission. Have all the mission-specific or unique requirements been effectively communicated to personnel? Have any additional skills required been developed?
  - Location. Have personnel been adequately trained to perform in the specific geographic location and in the specific facility where they are located?
  - Conditions. Have personnel been adequately trained and/or conditioned to perform in the environmental and work conditions unique to their workplace?
  - Organization. Have personnel been adequately trained on any procedures or job requirements unique to their specific organization.
  - Qualifications. Do personnel meet all local qualification requirements for the work being done under the conditions and at the place involved?
  - Other. Are personnel also trained in performing duties and tasks that may be assigned or required in addition to primary job assignments? Are other factors affecting the training and education required to perform specific tasks properly?

Assignment

Are the people that are authorized actually assigned to the organization?

- Number. Is the actual number of people assigned equal to the number authorized?
- Experience. Are the people assigned the authorized grade and do they have the total experience and time in grade necessary to perform work tasks in the existing environment?

Facilities/Hardware

Are the right tools available to perform the mission?

Design

Are the facilities and hardware designed to meet mission requirements under mission conditions?

- Safety precedence sequence. Was every effort made to eliminate recognized hazards by redesign or engineering? Were efforts then made to use physical guards and barriers to separate potential hazards from potential targets? Were warning devices then considered? Were procedures and training looked to as controls only after all other alternatives had been eliminated? Were any residual risks identified, analyzed, and accepted by top management?
- Designer qualifications. Do designers have the formal education and experience necessary to design high-quality, cost-effective, mission-oriented equipment and facilities?
- System safety effort. Were system safety analyses required? Were they appropriate? Were system safety recommendations included in the design?
- Safety analysis. Was a thorough safety analysis performed? Was a priority problem list developed? Was the safety analysis performed early in the design process, and were safety recommendations included in the design?
- User input. Is end user input solicited and used in the initial design effort, and are end users employed to review and validate designs?
- Budget. Are sufficient funds available for the project?
- Technical information. Do designers have and use properly timely accurate technical information?
  - Knowledge. Is the scientific and engineering data base sufficient to provide the technical knowledge required?
  - Communication. Is existing knowledge effectively communicated to designers?
  - Analysis. Is available technical knowledge analyzed and used properly?

- Other. Do other factors preclude the use of good, accurate, timely technical information?
- Criteria. Were the design criteria adequate to ensure a high-quality, cost-effective, mission-oriented product?
  - Performance. Were the performance criteria sufficient to ensure that the product will meet mission requirements?
  - Safety. Were safety, fire protection, and health requirements clearly described? Were they consistent with codes, standards, and regulations? Do they incorporate state-of-the-art protection methods?
  - Reliability. Are reliability requirements clearly stated? Are they sufficient to ensure a product that will meet mission requirements and perform with an acceptable, realistic level of reliability under mission conditions?
  - Maintainability. Do designs clearly state maintainability requirements? Are they sufficient to ensure that products can be maintained in a timely, cost-effective manner using available personnel and equipment under mission conditions?
- Human factors. Was the facility hardware designed to minimize human errors in operations and maintenance activities?
  - Stereotypes. Is the design consistent with learned and natural stereotypes? Do things operate the way they are expected to operate?
  - Ergonomics. Do things physically fit the people who are going to be using or operating them? Have the size, shape, and location of controls and instruments been carefully considered?
  - Human-machine. Does the system have machines doing tasks best performed by people? Are provisions made for people backing up machines and vice versa?
  - Misuse analysis. Has careful consideration been given to the way items can be used improperly and the possible effects of misuse? Is the system designed to be as "idiot-proof" as practical?

Construction/Fabrication

Were the facilities/hardware constructed or made as designed?

- Specifications. Were all design criteria effectively translated into construction/fabrication specifications and clearly communicated to contractors (or other production personnel), and did end products meet all specifications?
  - Performance. Does the end item do what it is supposed to do under the specified conditions?
  - Safety. Are the safety, health, and fire protection requirements met?
  - Reliability. Does the product meet reliability requirements and specifications?

- Maintainability. Can the end item be maintained as specified?
- Procurement process. Was the procurement handled properly?
- RFP. Was the request for proposal clearly written, and did it contain the necessary information and requirements to provide contractors with the information needed to submit a good bid?
  - Technical proposal. Were the requirements adequate to ensure a high-quality product that meets specifications? Were the requirements realistic for the scope and level of the project? Does the format and evaluation criteria ensure a fair, objective evaluation process?
  - Cost proposal. Are the requirements clearly written? Does the cost proposal information provide a basis of understanding in sufficient but not excessive detail?
  - Technical evaluation. Was the technical evaluation done objectively by qualified evaluators? Was the evaluation consistent with the evaluation factors listed in the RFP?
- Contract award. Was the contract awarded consistent with terms of the RFP? Were provisions made for objectively awarding to the "best value" or "best advantage to the buyer," not just the "low bidder"?
- Contract administrator. Was the contract administered fairly and objectively? Was the contractor required to perform as specified in the RFP and in accordance with the contractor's proposal? Were any changes managed properly and fairly (to the buyer and the contractor)?

Testing

Was the facility/hardware approximately tested to ensure that design criteria and construction/fabrication specifications were met?

- Timely. Was the testing done at the appropriate stage(s) of the life cycle? Were deficiencies detected early enough to make cost-effective fixes?
- Test criteria. Does the testing effort adequately measure and evaluate the performance, safety, reliability, and maintainability of the end item under specified conditions?
- Test design. Does the testing effort adequately duplicate real world conditions? Are safety, health, and fire protection safeguards incorporated into the test?
- Analysis. Were test results analyzed and evaluated properly? Were results fed back and appropriate "fixes" made?
- Validation. Were test results repeated and/or cross-checked by other methods to make sure that test data were accurate, relevant, and reliable?

Inspection

Are there adequate provisions for monitoring the status of the facility/hardware?

- Initial. Are provisions made for conducting a comprehensive acceptance inspection to accurately determine product status at the time of delivery?
- Ongoing. Are provisions made for conducting appropriate inspections periodically during the life cycle and/or at critical times before, during, and after operations?
- Plan. Does a comprehensive inspection plan clearly indicate when inspections are to be made, inspection responsibilities, qualification of inspectors, inspection criteria, and documentation/reporting requirements?
- Implementation. Is the inspection plan fully implemented? Are adequate resources (personnel, equipment, time, funding) available to sustain the program?

Maintenance

Is the facility/hardware maintained in such a manner as to maintain the original performance, safety, and reliability standards?

- Preventive. Are cost-effective services performed to extend the life and maintain the serviceability of the end item?
  - Plan. Does the preventive maintenance plan provide for timely, cost-effective services that protect facilities/hardware being maintained and also safeguard maintenance personnel and equipment?
  - Implementation. Is the plan followed?
- Upgrade. Are adequate provisions made in the maintenance operation to accomplish upgrades as required by manufacturers, major commands, or regulations?
- Corrective. Are broken things fixed rapidly and correctly?
  - Resources. Are adequate resources available to conduct unscheduled repairs?
  - Responsive. Are unscheduled repairs completed in a timely manner?

Change Control

When facilities or hardware are modified, are the changes engineered and controlled with the same care as a new acquisition?

- Environmental impact. Was the effect on the environment adequately considered when evaluating changes, and were the effects of environmental conditions adequately considered when proposing changes?

- Life cycle analysis. How will the change affect the life cycle of the end item? If the life cycle is extended, are all systems/subsystems adequate to support the extended life? Is a change analysis performed?
- Replacement plan. Does a clear, well-written, and properly distributed plan indicate when and how replacement of components, subsystems, and end items should be accomplished? Does it contain sound replacement criteria?
- Upgrades. When decisions are made to upgrade the performance, safety, reliability, or maintainability of an item, is a change analysis made to determine the effect of the upgrade on the other factors?
- Disposal plan. Do changes affect the eventual plans for disposing of the end item? If so, have the appropriate changes been made in the disposal plan?
- Modification plans. Are all changes or modifications carefully planned to ensure that all appropriate counterchanges are made? Is a comprehensive change analysis part of the modification plan?
  - Maintenance plans. Are maintenance plans, procedures, and support equipment modified as required to match modifications made to the facility or hardware? Are maintenance personnel trained on the new procedures?
  - Inspection plans. Are operating plans and procedures modified?
  - Human factors. Are modifications consistent with learned and natural stereotypes? Do things operate the way they are expected to operate?
    - Stereotypes. Are the modifications consistent with learned and natural stereotypes? Do things operate the way they are expected to operate?
    - Ergonomics. Do things physically fit the people who are going to be using or operating them? Have the size, shape, and location of the controls and instruments been carefully considered?
    - Human-machine interface. Do modifications adversely affect the way the people and the machines in the system work together?
    - Misuse analysis. Are modifications as "idiot-proof" as practical? How can modified facilities or hardware be misused?

REVIEW QUESTIONS

1. What are the types of system safety analysis for which PET is best suited?
2. What are two other uses of PET?
3. What is the recommended color coding for the PET chart?
4. Explain how PET is self-tailoring.
5. Explain the relationships among PET, mini-MORT, MORT, and COMPAS.

REFERENCES

Buys, J. R. 1977. *Standardization Guide for Construction and Use of MORT-Type Analytical Trees.* ERDA 76-45/8; SSDC-8. Idaho Falls, ID: Energy Research and Development Administration.

Johnson, William G. 1973. *MORT, the Management Oversight and Risk Tree.* Washington, DC: U.S. Atomic Energy Commission.

Johnson, William G. 1980. *MORT Safety Assurance Systems.* New York: Marcel Dekker.

CHAPTER 17

Change Analysis

Change is a necessary ingredient of progress. It can be a positive factor that improves the effectiveness, efficiency, and/or safety of an organization or operation. However, change is almost always a causal factor in accidents, many times a very significant causal factor. Whether change is a friend or foe depends largely upon the manner in which changes are planned, managed, and controlled.

Change analysis is one of the techniques associated with the Department of Energy's management oversight and risk tree (MORT) approach to system safety. Unlike the MORT chart itself or some of the other tools and techniques associated with the MORT program, change analysis is a very simple, straightforward process that is relatively quick and easy to learn and to apply.

Well-designed and engineered systems can rapidly become ineffective, inefficient, and/or hazardous without effective change controls.

Change analysis is a technique for evaluating changes and determining the need for counterchanges to keep the system in balance.

Change analysis has applications in both accident prevention and accident investigation. It is recommended for both.

PURPOSE

The purpose of change analysis is to list and examine all changes systematically, determine the significance or impact of the changes, and recommend appropriate counterchanges to ensure that the safety of the operation is not adversely effected by the changes.

Change analysis, as a preventive tool, provides an excellent method for conducting reviews. Thus, a change analysis should be performed as part of the review process at each review point in the system safety effort.

Change analysis is also a very important accident analysis tool and should be performed as part of any accident investigation. It is particularly well suited

System Safety for the 21st Century: The Updated and Revised Edition of System Safety 2000, by Richard A. Stephans
ISBN 0-471-44454-5   Copyright © 2004 John Wiley & Sons, Inc.

for finding quick answers and identifying obscure direct causes. Change analysis can also explain situations in which seemingly rational individuals revert to "Keystone Cop" behavior. It is normally the result of being overwhelmed by change.

INPUT REQUIREMENTS

In order to perform a change analysis, a detailed understanding of the system before and after the changes is required. Standard project description documents, including analytical trees, narrative descriptions, drawings, block diagrams, blueprints, schematics, and, if applicable, operating and maintenance procedures, organizational structures, job descriptions, and personnel qualifications, can all aid in understanding the project.

GENERAL APPROACH

The general approach to change analysis, whether for review or accident analysis purposes, is extremely simple (Fig. 17-1).

The present (or accident-free) situation is compared with the proposed (or accident) situation. All changes (that is, the differences between the present situation and the comparable situation) are systematically listed. Then each change or difference is carefully analyzed for possible significance.

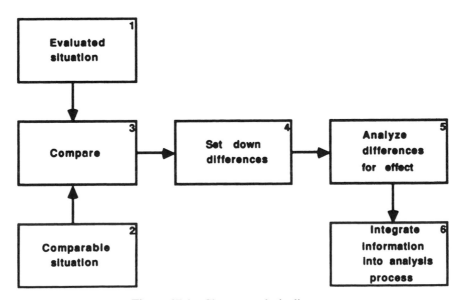

Figure 17-1 Change analysis diagram.

Finally, where applicable, counterchanges are recommended to ensure that the changes identified do not degrade the overall safety of the project or operation (or to ensure that changes that were causal factors to an accident are adequately controlled with counterchanges).

INSTRUCTIONS

First, determine the two situations to be compared. When change analysis is used as a review tool, this determination is quite simple; the new or proposed system or situation is the evaluated situation to be compared with the old or existing system or comparable situation.

For accident analysis or investigation applications, obviously the accident situation is to be compared to some comparable situation. There are three basic alternatives for this comparable situation. First, it can be the same operation or organization at an earlier point in time. For example, compare the personnel, procedures, plant and hardware, environment, tasks, and other factors at the time of the accident with these same factors as they were one hour (or day or week or year) before the accident. The selection of the comparable point in time depends upon the nature of the operation and the speed at which significant changes occurred.

A second choice for a comparable situation is to compare the accident situation with a similar accident-free operation. For example, two similar plants producing the same products from the same raw materials in the same environment could be compared, or two military units with the same mission, same organization, same type of equipment, vehicles, or aircraft, subject to the same regulations, and in the same physical and command environments (for example, side-by-side tank battalions or fighter squadrons) can be compared.

A third choice—probably the most frequently used—is to compare the accident situation with an idealized situation; that is, compare the actual events with the events that should have occurred if everything was operating correctly. Compare the personnel involved with authorized, full-strength, fully trained, ideal cadres; compare the equipment present with the equipment that should have been present; and compare the environment with a "normal" environment.

The selection of an appropriate comparable situation is probably the most difficult step in performing a change analysis, but even this step is quite simple if the change analysis is part of a review of proposed changes.

The next step is to collect sufficient information about the proposed (or accident) situation and the comparable situation to understand fully the personnel, procedures, plant and hardware, environment, and tasks associated with each. If the change analysis is part of a design review, this step may entail only reviewing relevant drawings and project documentation. However, an accident analysis may require extensive collection and analysis of the "people evidence," "physical evidence," and "paper evidence" associated with any in-depth accident investigation.

The next step is carefully listing *all* the differences between the proposed (or accident) situation and the appropriate comparable situation. The key is to search for all changes or differences and to avoid evaluation or analysis of the significance of the changes until all changes have been identified and listed.

After all changes have been listed, each change is individually evaluated and analyzed. The significance of each change is noted, and, where appropriate, controls or counterchanges are recommended to ensure that the changes listed do not adversely affect the project.

A summary of significant changes, to include a discussion of potential impact and recommended changes, is prepared and fed back into the system as part of the review process, accident investigation, or other appropriate protocol.

If not already established, a tracking system should be initiated to ensure adequate follow-up.

Worksheet

Heading. Fill in appropriate blocks (Fig. 17-2).

Column 1—Evaluated Situation. Use the key words listed to compare carefully every aspect of the proposed (or accident) situation with the comparable (present or accident-free) situation.

For each factor that is changed or different, list that factor as it applies to the proposed (or accident) situation in column 1.

Column 2—Comparable Situation. For each factor listed in column 1, list the comparable factor for the present or accident-free situation in column 2.

Column 3—Significance. In column 3, discuss the significance or impact of the differences between the factors listed in column 1 and the corresponding factors listed in column 2. (If none, so state.)

Column 4—Comments. In column 4 discuss any hazards or potential hazards associated with the change, and recommend appropriate controls or counterchanges to restore or improve the safety of the operation.

A Sample Change Analysis

An example of how change analysis can be used to find an obscure direct cause relatively quickly is contained in a story frequently told by instructors from the Department of Energy's System Safety Development Center.

Personnel working at the Idaho National Engineering Laboratory (INEL) are transported approximately forty miles from Idaho Falls to the INEL site

Factors	Evaluated Situation	Comparable Situation	Differences/ Distinctions	Significance
NOTE: CONSIDER THE FOLLOWING FACTORS BUT DO NOT ATTEMPT TO "MATCH UP" OR ALIGN FACTORS WITH INFORMATION IN COLUMNS				
What Object(s) Energy Defects Protective Devices				
Where On the Object In the Process Place				
When In Time In the Process				
Who Operator Fellow Worker Supervisor Others				
Task Goal Procedure Quality				
Working Conditions Environmental Overtime Schedule Delays				
Trigger Event				
Managerial Controls Control Chain Hazard Analysis Monitoring Risk Review				

Subject _____

Figure 17-2 Change analysis worksheet.

by buses. One morning on the way to the site, several buses experienced brake failure. Fortunately, no accidents resulted, but the cause for the brake failure had to be determined before the return trip at the end of the workday.

The safety professionals at the System Safety Development Center were contacted for assistance in determining what had happened and in determining the appropriate course of action concerning the use of the buses at the end of the day to transport the work force back home.

The supervisor of the bus maintenance organization was telephoned and was able to offer no explanation. None of the failed brake systems had any mechanical damage.

When asked specifically about any changes to the bus brake systems, the maintenance supervisor stated that, in preparation for winter, antifreeze had recently been added to the brake systems; the temperature on the day of the mishap was, in fact, well below freezing. This change, however, should have prevented rather than caused the problem.

The maintenance supervisor was then asked to outline any other recent changes to the brake systems. After a long silence, he replied, "That's it!" A couple of weeks earlier, in response to suspected water contamination in some systems, a desiccant had been added to the brake systems to absorb any water. The desiccant had a higher affinity for antifreeze than for water, so that when the antifreeze was added, it was absorbed and the water released back into the system. The water then froze, causing the brake failures.

Because the temperature in the afternoon was above freezing, the buses could safely return to town, where the suspect systems could be drained and refilled overnight.

By concentrating on changes, the problem was solved rapidly and accurately.

REVIEW QUESTIONS

1. When is change analysis used?
2. At what point in the analysis is the significance of identified changes considered?
3. When using change analysis as a review tool, what should be used as the "evaluated situation" and what should be used as the "comparable situation"?
4. When using change analysis for accident investigation, the accident situation is the evaluated situation. What are the three choices for the comparable situation?
5. In addition to general comments, what else belongs in column 4 (Comments) on the change analysis worksheet?

REFERENCES

Johnson, William G. 1973. *MORT, the Management Oversight and Risk Tree.* Washington, DC: U.S. Atomic Energy Commission.

Johnson, William G. 1980. *MORT Safety Assurance Systems.* New York: Marcel Dekker.

CHAPTER 18

Management Oversight and Risk Tree

Bill Johnson developed MORT or the management oversight and risk tree in the 1970s as part of his overall systems safety approach and systems safety effort for the Department of Energy, at that time the Atomic Energy Commission. The MORT chart contains approximately 1,500 base items arranged into a large and rather complex fault tree that is primarily used for accident investigation.

The symbols used on the MORT chart are basically those used for other analytical trees (Chapter 10) and fault tree analysis (Chapter 15). They include the rectangle as the general event symbol, the circle as the base event symbol, the diamond as an undeveloped terminal event, the *and* gate, the *or* gate, and the ellipse as a constraint symbol (Figs. 18-1 and 18-2).

In addition, Bill Johnson used a scroll as a "normally expected event" and an oval as a "satisfactory" event. The normally expected event distinguishes events that are typically a part of any system, such as change and normal variability. The satisfactory event describes events that may be accident causal factors but are a necessary part of the operation, like "functional" (part of the system) people or objects in the energy channel. Also, in addition to using the traditional transfer symbol (a triangle), the MORT chart includes capital letters as drafting breaks and small ellipses as risk transfers (Fig. 18-3).

Risk transfers are found in two places. They are found next to undeveloped terminal events, and they are found near the top of the chart as assumed risks. The implied logic is that barriers, controls, or evasive actions are not possible or practical in those areas where there are undeveloped terminal events. If activity is continued under these circumstances, then some risk is assumed or accepted. The small assumed risk transfers are used to denote a transfer up to the small ovals under the assumed risk portion of the MORT chart.

Abbreviations used on the MORT chart include *LTA*, the most frequently used, which stands for "less than adequate"; *DN* for "did not"; *FT* for "failed to"; *HAP* for "hazard analysis process"; *JSA* for "job safety analysis"; and finally *CS&R* for "code standards and regulations".

System Safety for the 21st Century: The Updated and Revised Edition of System Safety 2000, by Richard A. Stephans
ISBN 0-471-44454-5   Copyright © 2004 John Wiley & Sons, Inc.

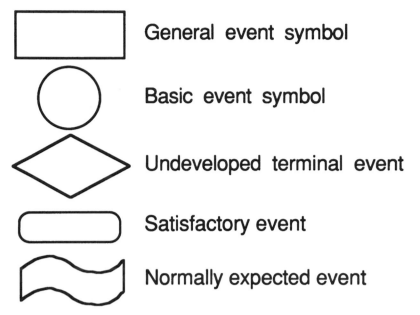

General event symbol

Basic event symbol

Undeveloped terminal event

Satisfactory event

Normally expected event

Figure 18-1   Event Symbols. The General Event symbol, Basic Event symbol, and Undeveloped Terminal event symbol are the same as commonly used for other fault trees. The use of the stretched circle for a Satisfactory Event and the scroll for a Normally Expected Event are unique to the MORT chart.

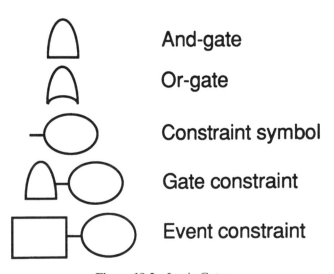

And-gate

Or-gate

Constraint symbol

Gate constraint

Event constraint

Figure 18-2   Logic Gates.

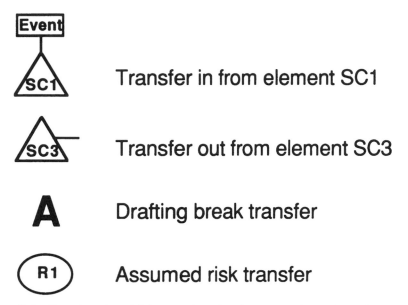

Transfer in from element SC1

Transfer out from element SC3

Drafting break transfer

Assumed risk transfer

Figure 18-3 Transfers. In addition to the triangle commonly used as the transfer symbol for analytical trees, the MORT chart also uses large capital letters as "drafting break" transfers and small ellipses as transfer symbols for assumed risks.

The event symbols, gates, transfers, and abbreviations are all shown on the legend in the lower left corner of the large MORT chart.

A key MORT definition is that of *accepted* or *assumed risk*. An assumed risk is a very specific risk that has been identified, analyzed, quantified to the maximum practical degree, and accepted by the appropriate level of management after proper evaluation.

This particular definition has three fairly subjective parts. The first is "quantified to the maximum practical degree," which obviously may at times be difficult to determine accurately. A second subjective part involves acceptance by the "right" level of management, which is seldom determined to be too high. The errors are generally in accepting risk at too low a level. The third subjective bit of wording is "after proper evaluation." It provides an adequate working definition for accepted or assumed risks; however, sometimes pre-accident assumed risks are considered to be "oversights and omissions" when reevaluated after an accident.

The primary advantage of the MORT chart for accident investigation is that it aids the investigator in making an in-depth investigation and identifying the root causes of the accident. It does this by providing a systematic means of evaluating both the specific control factors and the management control factors that caused or contributed to the accident. The largest single disadvantage to the MORT approach is that it is time-consuming and tedious to learn. It is perceived as overkill for many accidents.

The analogy can be made to using a word processor. Early attempts to use a computer to produce a letter can be very frustrating. Learning how to set up and turn on the hardware, boot the software, enter the appropriate commands, and finally print out the product can take many times longer than simply sitting down at a typewriter and typing a letter. Once the investment in time and energy has been made and the computer has been mastered, however, it really is a gigantic timesaver, and the letter writer usually has no desire to use a typewriter ever again.

The MORT chart is very much that way. It is extremely time-consuming and tedious when the MORT analysis is first attempted. If the investigator is diligent, however, and learns how to use it, the MORT chart then begins to be truly a time-saving accident analysis device. Using a comprehensive checklist—albeit complex—to perform a systematic, in-depth investigation is easier than starting with only a blank sheet of paper.

The analogy and logic are sound, and MORT analysis with the large MORT chart is recommended for those individuals who routinely conduct investigations of major accidents and spend a significant portion of their time serving on accident investigation boards.

However, only a small fraction of accidents require investigation by formal boards of investigation. Most organizations use some sort of hierarchy so that minor mishaps are investigated by first-line supervision, accidents with moderate consequences (or potential consequences) are investigated by the safety organization (with formal or informal input by supervision), and only the most serious accidents, in terms of actual or potential losses, are investigated by formal boards of investigation. The classes or types of accident (typically *A*, *B*, and *C*) should be clearly defined in terms of severity and/or numbers of injuries, dollar losses, programmatic impact, and other parameters and should automatically trigger the appropriate level of investigation. Because membership on boards of investigation is normally tailored to the accident and the situation, few individuals, even safety professionals, spend enough time performing major accident investigations to develop and maintain proficiency with the MORT chart. (Developing word processing skills may not be practical for those who type only one letter a year.)

Yet, the MORT approach is extremely valuable in ensuring that the root causes of all accidents are identified and that corrective actions are taken.

In response to complaints that the MORT chart is overkill for may situations and that MORT analysis, using the MORT chart, is too time-consuming to teach to supervisors and to accident board members, I developed the mini-MORT chart in 1980 and used it to introduce supervisors to MORT concepts. During the mid-1980s mini-MORT was introduced to the U.S. Air Force ground safety community as an alternative tool for single-investigator accidents and later in the decade was picked up by the System. Safety Development Center and included in the Department of Energy MORT program.

The mini-MORT chart is basically the full-size MORT chart with the bottom tier removed to reduce the number of events to be evaluated from approximately 1,500 to 150. This simplification also makes possible eliminating all transfers from the chart. Additionally, the seldom-used event symbols like the scroll and the stretched circle are replaced with the more familiar circles and rectangles. (I also developed the PET chart as another alternative to the MORT and the mini-MORT; see Chapter 16.)

PURPOSE OF MORT AND MINI-MORT

The purpose of MORT analysis is to provide a systematic tool to aid in planning, organizing, and conducting an in-depth, comprehensive accident investigation (or inspection, audit, or appraisal) to identify those specific control factors and management system factors that are less than adequate and need to be corrected to prevent recurrence of the accident (or to prevent other undesired events).

The purposes of the mini-MORT chart are to aid in teaching MORT concepts and, for relatively minor accidents, to serve as a tool for performing MORT analysis on a reduced scale.

The MORT analysis is primarily an accident analysis tool, even though it can also be used to inspect and analyze the system with respect to other undesired events.

INPUT REQUIREMENTS

Input requirements for MORT analysis, like those for any other detailed, comprehensive analysis, are extensive. Detailed information about the hardware and facilities, environment, procedures, and personnel directly involved is required for full evaluation of the specific control factors, and additional information about the management system is required to evaluate the policies, procedures, implementation plans, risk assessment program, and other upstream factors. Normally, a combination of interviews, physical inspections, and reviews of many procedures and project documents is required. The MORT chart and/or mini-MORT chart can serve as a planning and organizational tool for the collection of evidence and other relevant information.

GENERAL APPROACH

The MORT analysis effort begins as soon as the accident occurs and the accident board is notified. The MORT analysis is usually performed by the trained investigator on the board (or an advisor to the board). The MORT (or mini-MORT) chart is used as a working tool to aid in gathering information and storing information as it is gathered.

A color coding system is used with the chart (Fig. 18-4). Any event or factor

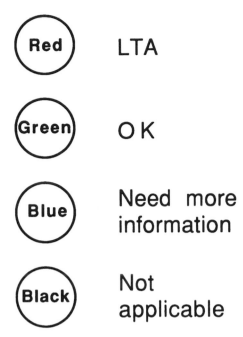

Red — LTA

Green — O K

Blue — Need more information

Black — Not applicable

Figure 18-4 Color coding. These colors are generally used on the MORT or Mini-MORT chart when performing MORT Analysis. They are also used for the PET chart.

found to be less than adequate (ISA) is colored red and addressed in the accident report with appropriate recommendations to correct the LTA condition. Obviously, any finding of LTA must be documented and supported with facts, and red is therefore used judiciously.

Chart events that are evaluated and found to be adequate are colored green. Because accident investigations frequently present an excellent opportunity to correct system deficiencies, green also ought to be used judiciously and supported by credible evidence.

The MORT chart is designed to encompass any accident situation, and not all parts of the chart are relevant to all accidents. A particular block or branch of the chart that is not applicable is color coded black (or sometimes simply crossed out). Again, if a block or branch is incorrectly colored black, the chance to make needed corrections is lost.

The final color used in MORT analysis is blue, which indicates that the block has been examined but that insufficient evidence or information is available to evaluate the block. A blue mark indicates that more evidence is to be collected in this area. Ideally, all blue blocks are replaced by one of the other colors when the investigation is complete; realistically, eliminating blue blocks may not always be possible.

The general method for working through the chart is from known to unknown, initially entering information as it becomes available and then using the incomplete portions of the chart to guide the investigation.

Specific control factors

Management system factors

Duality

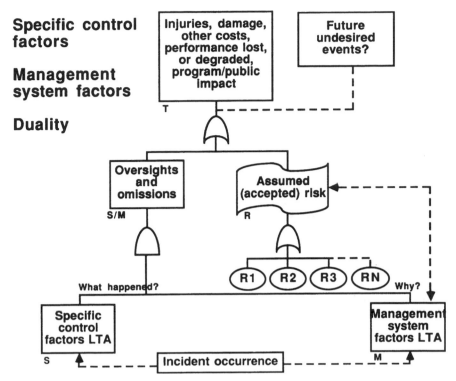

Figure 18-5 Top Events. The top tiers of the MORT chart indicate that the losses resulting from the accident being investigated result from oversights and omissions or assumed risks or both.

The top of the chart can generally be addressed very early in the investigation (Fig. 18-5).

Keep in mind that the chart is only a working tool to be used during the investigation to write the report and then to be discarded or filed with the background documentation; therefore, neatness does not count, and many analysts make notes directly on the chart. For example, a description of the accident specifying the losses is entered near the top block. (If the analysis is being performed to examine conditions that could lead to future undesired events, the specific undesired event is described.)

The oversights and omissions and assumed (accepted) risks tier of the tree is relatively academic and generally adds to the analysis. A formal accident investigation is unlikely to be performed on losses that result from assumed risks only. Losses from assumed risks are normally those from earthquakes, tornados, hurricanes, and other acts of nature. (Of course, exceptions exist, as do situations in which certain losses due to even remote acts of nature are not acceptable.)

For most investigations, however, analysis begins at the specific control factors and management control factors blocks. Generally, the specific control

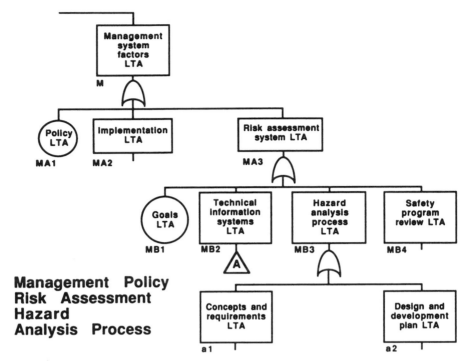

Management Policy
Risk Assessment
Hazard
Analysis Process

Figure 18-6 Management system factors. The alphanumeric notations near the lower left of each event symbol were originally used as a cross reference to the MORT User's Manual. Later versions of the MORT chart now use page numbers for cross reference purposes.

factors side of the tree tends to answer questions about what happened and addresses the accident documentation requirements that are part of all formal investigations. The MORT approach tends to answer these questions in more detail than many traditional methods. What tends to make MORT effective in identifying root causes, however, is the management system factors side of the chart, where questions of why are asked that ultimately lead to the upstream conditions and factors that need to be corrected to prevent accident recurrence (Fig. 18-6).

Management system problems can occur in three primary areas. First, the policy can be less than adequate. Actually, most organizations have carefully worded safety policy statements signed by the appropriate manager. Problems are much more likely to occur in one or both of the other areas.

Even the best policy is of little value unless it is implemented. Organizations many times fail to do what is outlined in the policy statement. Problems in this branch can frequently prevent "paper" programs from being effective.

Another major management system factor is the risk assessment system. If the risks are not identified, evaluated, and reported accurately to management,

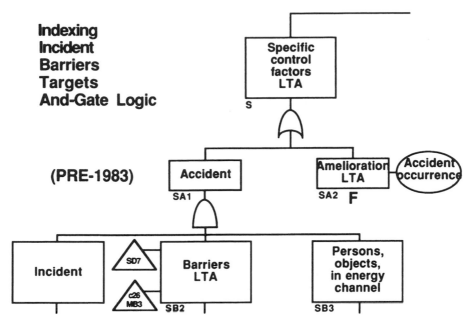

**Indexing
Incident
Barriers
Targets
And-Gate Logic**

(PRE-1983)

Figure 18-7   Special control factors (Pre-1983).

taking the appropriate steps to control or eliminate hazards or risks will be difficult. Oversights or omissions in this area also are frequent causal or contributing factors that must be corrected to prevent accident recurrence. Of course, the why questions on the management system factors side of the chart cannot be answered until the what happened questions on the specific control factors side of the chart have been addressed.

During the two decades since Bill Johnson introduced the original MORT chart, several modifications have been made by the Systems Safety Development Center. Prior to 1983, the first-tier events under *accident* on the specific control factors side of the chart were as depicted in Figure 18-7.

This configuration is consistent with the accident triangle (Fig. 18-8) and the definitions of *incident* as an unwanted energy flow and of *accident* as an unwanted energy flow that results in adverse consequences.

The energy flow and barrier analysis (ETBA) technique described in Chapter 13 evolved from this part of the MORT chart, and the mini-MORT chart was developed from this configuration. This version of the chart had a block (not shown) under the incident branch that indicated that in cases of multiple energy flows the analysis process should be repeated for each energy flow and barrier failure that led up to the accident (Fig. 18-9).

In 1983, the first tier under *accident* was reconfigured as shown in Figure 18-10. The most significant change was adding "potentially harmful environmental conditions" to the chart. It was probably at least partially motivated

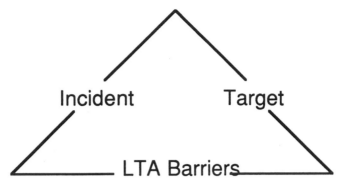

Figure 18-8 Accident triangle. The "incident", "LTA barriers", and vulnerable "target" concept is the basis for Energy Trace and Barrier Analysis (ETBA) described in Chapter 13.

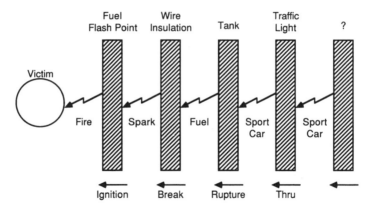

Figure 18-9 Unwanted energy flow. Many accidents are the result of a series of unwanted energy flows, barrier failures, and vulnerable targets.

by the interaction that SSDC was having at NASA during this period, because potentially harmful environmental conditions associated with space travel and training for space travel are very significant NASA concerns. Also, the case study used as the primary practical exercise in teaching MORT-based accident investigation is about a near-asphyxiation in an argon atmosphere. Treating the inert argon atmosphere as an "unwanted energy flow" was always somewhat awkward, and the 1983 changes made the case study fit the chart more readily.

The "event and energy flow leading to accident-incident" block replaced the "precedent energy flow" block from the previous chart and made the concept of multiple energy flows clearer and easier to address on the chart.

"Amelioration" continues to appear on the same tier as the accident. *Amelioration* refers to postaccident actions such as medical services, fire

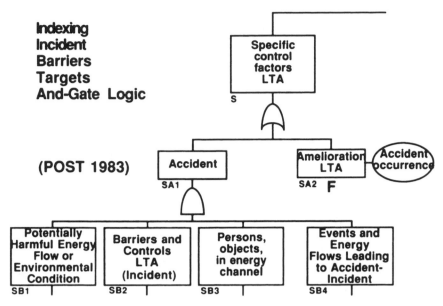

Figure 18-10   Specific control factors (post-1983). These changes made it easier to include potentially harmful environmental conditions and to understand multiple energy flows, but do not significantly alter the analysis process.

fighting, rescue efforts, and public relations. The postaccident response frequently has a very significant impact on the magnitude of the losses. A very good response can minimize losses, whereas a poor response may produce losses much greater than those caused by the original accident.

For analysis purposes, the changes under the accident block make little difference. The three main branches on the specific control factors side continue to be the contributory events that allowed the unwanted energy flow (or hazardous environment), the LTA barriers, and the targets in the energy path (or vulnerable to the hazardous environment).

The items to evaluate to determine the adequacy of the controls designed to prevent unwanted energy flows and exposures to hazardous environments include the technical information system, maintenance and inspection programs, direct supervision and services provided by higher levels of supervision, and the "functional operability" of facilities (Fig. 18-11).

Barrier analysis includes determining whether or not the barrier failed, if a barrier was provided and used. If a barrier was provided but not used, the analysis then seeks to identify the task performance errors which caused the barrier not to be used (Fig. 18-12). This process is used to evaluate barriers on the energy source, between the energy source and the persons and/or objects to be protected, or on the target person or object. The same process is also used if the barrier evaluated consists of separating the potential targets from potentially harmful energy flows or environments by time or space.

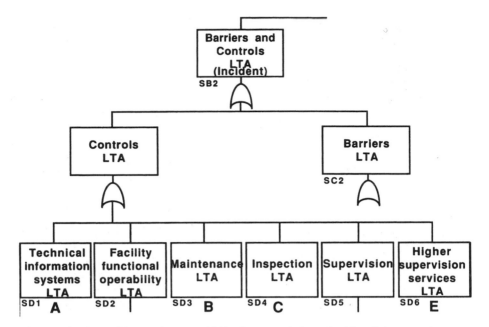

Figure 18-11   Incident. The post-1983 charts redefine "incident" but continue to analyze the same contributing factors as earlier charts.

In analyzing the persons or objects in the energy channel, first the analyst determines whether the persons or objects were needed at the specific time and place where they were vulnerable as targets; that is, were they a necessary part of the operation (functional) or were they merely in the wrong place at the wrong time as visitors, spectators, passers-by, intruders, or, in the case of objects, merely stored in a vulnerable location unnecessarily (nonfunctional)? For both groups, the adequacy of controls is an item to be investigated. For the functional group, an evaluation of administrative controls and possible evasive, action is also required (Fig. 18-14).

Figures 18-5, 18-6, and 18-10 through 18-13 represent the major upper tiers of the MORT chart but are less than 1% of the total events to be analyzed. The MORT chart is tedious to use not only because of its size but also because of the use of multiple transfers and transfers within transfers throughout the chart (Fig. 18-14).

Thus MORT analysis, using the full-size MORT chart, is recommended primarily for major accidents and for use by a trained investigator who has attended at least forty hours of instruction in MORT (preferably one of the SSDC courses or the USAF advanced occupational safety course).

Whether using the MORT chart or the mini-MORT chart to perform MORT analysis (or using the PET chart for PET analysis), the general approach is to use the chart as a working paper, begin as early as possible, add information as it is obtained until the chart is complete, and then use the com-

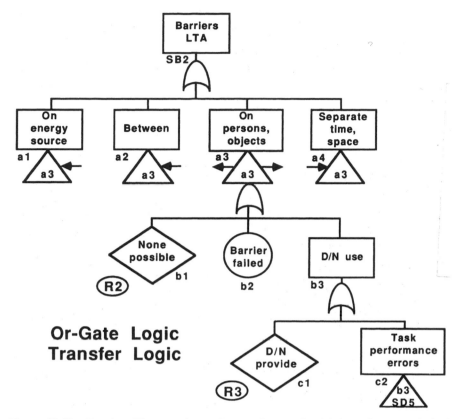

Figure 18-12   Barriers. The use of transfers tends to make this branch seem relatively simple. However, there are almost 400 events to evaluate under the Barriers LTA branch.

pleted chart, the event and causal factors chart (see Chapter 19), and change analysis information (see Chapter 17) to write the formal accident report. All of the major analytical tools (MORT, mini-MORT, PET, event and causal factors, and change analysis) are developed and used together, cross-checking and interrelating information as the investigation and analysis continue.

INSTRUCTIONS

The following instructions are for performing MORT analysis using the mini-MORT chart. The mini-MORT chart was chosen because it is easier to use and to learn and is probably better suited to all except catastrophic accidents.

The full-size MORT chart and a manual for it (SSDC-4, Revision 2, *MORT User's Manual*) are available from the Department of Energy's System Safety

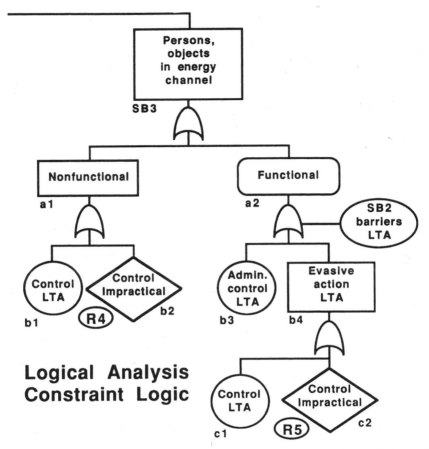

Figure 18-13 Targets. This branch addresses the persons and/or objects which are the targets of the unwanted energy flow (or harmful environmental condition).

Development Center, EG&G Idaho, Inc., P.O. Box 1625, Idaho Falls, Idaho, 83415.

The MORT chart and user's manual are also reproduced in *MORT Safety Assurance Systems*, by William G. Johnson (published by Marcel Dekker, Inc., New York, 1980, in cooperation with the National Safety Council; available from the NSC, 444 North Michigan Avenue, Chicago, Illinois, 60611). Working with the 22-inch by 34-inch chart is much easier than working from the dissections of the chart found in *MORT Safety Assurance Systems*.

System Safety 2000 is the only known source of instructions for using the mini-MORT chart.

Following the general approach previously described, use the following questions and comments to color code the chart. Make sure that adequate notes are made to substantiate the evaluations made, especially red or less

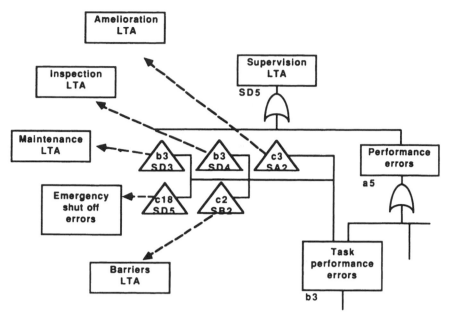

Figure 18-14 Multiple Transfers. Many branches of the MORT chart need to be evaluated several times.

than adequate determinations. These notes, with the chart, are extremely useful for writing the final report.

The top tiers of the chart are academic and, for all practical purposes, can be colored red as soon as the investigation begins. Keep in mind that this analysis is not a coloring-book exercise. Colored dots or check marks are recommended, especially for the initial blue events, as their color must change before the analysis is complete. Not uncommonly, new evidence causes changes even in blocks initially colored red, green, or black.

Because red logically drives the chart—that is, a red or LTA base event causes the blocks above the *or* gates to be red—the analyst can relatively safely assume that if an accident investigation is in progress that oversights and omissions occurred and that something LTA is on both sides of the chart.

Looking at the chart from left to right and generally from top to bottom, the first branch to be analyzed, then, is "Amelioration" (Fig. 18-15).

Amelioration refers to postaccident actions, which generally play a large role in determining the magnitude of the total losses. Factors to be evaluated include

> *Rehabilitation:* How soon and how efficiently were operations back to normal? This topic includes the physical rehabilitation of injured personnel and the repair or replacement of hardware, software, utilities, and facilities. Were plans in place to aid in rehabilitation efforts?

PAGE 1

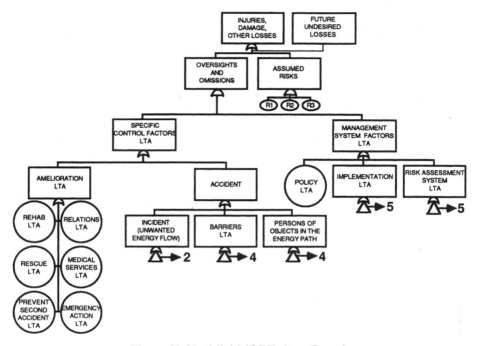

Figure 18-15   Mini-MORT chart, Page 1.

Were they adequate? Were they implemented? How could they be improved?

Relations: How were relations handled with the families of injured personnel? Were they notified in a prompt, accurate, professional manner. How were relations handled with the press, the community, higher organizational units, unions, clients, and any other people or organizations that may be affected by the accident? How were relations with insurance carriers and outside emergency response personnel? Was there a plan? Was it adequate? Was it implemented? How could it be improved?

Rescue: Did the appropriate emergency response organizations do their job properly to remove personnel and/or property from the accident environment? Were evacuation and rescue plans in place? Were they adequate? Were they implemented? How can they be improved?

Medical services: Was appropriate first aid administered on the scene? Was the medical response rapid and adequate? Were the injured handled and treated properly at the accident scene, in transit, and at the medical facility? Were emergency medical services adequately planned? Was the plan implemented? How could it be improved?

Prevent second accident: Were the losses limited to those that resulted from the initial unwanted energy flow? Were rescue or other emergency response personnel injured? Was unnecessary damage to property done during the rescue effort?

Emergency action: Did the system respond properly? Did alarms and reporting systems work? Did individual employees take the appropriate actions? Did emergency response personnel take the appropriate actions? Did management take the appropriate actions? Were adequate plans in place? Were they implemented? How could they be improved?

Assuming an accident did, in fact, occur, the next branches to be analyzed are those that may have contributed to the unwanted energy flow (Fig. 18-16). *Higher supervision services* are provided by levels of supervision between the

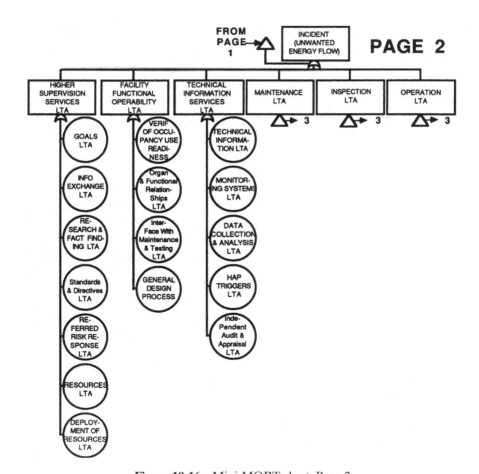

Figure 18-16   Mini-MORT chart, Page 2.

direct supervisor and "management." Obviously, they are going to vary from organization to organization. Typically, for example, if the direct supervisor is a foreman, "higher supervision" includes the general foreman and the superintendent, and the project manager or production manager is the first "management" tier. The investigator defines *higher supervision* as it applies to the accident situation.

Goals: Were the job goals and safety goals set by higher supervision realistic, obtainable but challenging, and consistent with organizational goals?

Information exchange: Was information about job tasks, safety requirements, hazards, rights and responsibilities, and other relevant information passed accurately and in a timely manner between direct supervision and management in both directions? Did middle levels of supervision shortstop, filter, or inappropriately alter communications? Did middle supervision interpret or supplement information as necessary?

Research and fact-finding: Did middle supervision search out and provide the information, including safety requirements, necessary for safe job performance?

Standards and directives: Did middle supervision maintain the appropriate standards and directives? Were they kept up to date and accessible to first-line supervisors and workers?

Referred risk response: Were supervisors responsive to situations in which hazards were reported to them? Did problems, concerns, or requests for assistance in analyzing or reviewing risks receive appropriate responses?

Resources: Did middle supervision ensure that the necessary personnel, tools, equipment, procedures, and time were available to do prescribed tasks?

Deployment of resources: Did middle supervision ensure that available resources were distributed fairly and consistent with job needs?

Facility functional operability refers to whether the facility in which the unwanted energy flow or accident occurred was fully operational and configured for the type of work being done. It is usually of greatest concern in very new facilities, where a desire to start operations prematurely may contribute to an unsafe condition, and in very old buildings, where insidious changes may have allowed a facility once suited to its operations to deteriorate to the point where it is no longer suitable. It is frequently a causal factor in situations where the purpose of the facility has been changed (for example, warehouses converted to offices, aircraft hangars converted to maintenance facilities, maintenance facilities designed for one type of vehicle now used for much larger vehicles with different types of engines). Specific items to evaluate include

Verification of occupancy use readiness: Was the facility ready for operations when operations began? Were maintenance and inspection programs in place? Had required personnel training been conducted on operation of facility systems? Had emergency evacuation procedures been established? Were all utilities operating properly? Was construction complete? Were safety, security, and communications systems in place?

Organizational functional relationships: Were responsibilities clearly defined? Did everyone understand clearly who owned what and where everyone belonged?

Interface with maintenance and testing: Were maintenance responsibilities clearly defined? Were lockout and tagout procedures in place and clearly understood by maintenance and operations personnel? Was an adequate plan in place to check out or test all the systems during start-up?

General design process: Is the overall design safe and efficient? Was a system safety effort in place for the facility? Was safety and user input incorporated into the design? Did the design process, including changes, provide for adequate review throughout the design and construction process?

Technical information services is concerned with whether the necessary information about the energy sources and the operation was known and communicated. Factors to evaluate include

Technical information: Was adequate, accurate information about the energy source known? Was it communicated effectively to all individuals needing the information? Were all available information sources used (including external data banks, consultants, studies, current research)?

Monitoring systems: Were the monitoring systems in use appropriate for the energy source? Were the monitoring systems operating properly? Was the information from the monitoring effort valid, useful, and communicated effectively? Was a mix of monitoring techniques employed?

Data collection and analysis: Was available information collected, distributed, and used properly? Was information collected, analyzed, and communicated in a timely manner? Was the analysis accurate and the results meaningful? Were data collected from different sources correlated and cross-checked?

HAP Triggers: Was an adequate plan in place to initiate the hazard analysis process (HAP) based on input from the monitoring system or from other sources of technical information?

Independent audit and appraisal: Were provisions made for objective "outside eyes" to perform audits and appraisals? Had independent audits and appraisals been performed? Recently? Were they adequate? Had the results been communicated and used properly?

The *maintenance* questions (Fig. 18-17) address whether an adequate maintenance program was in place. Factors to be evaluated include

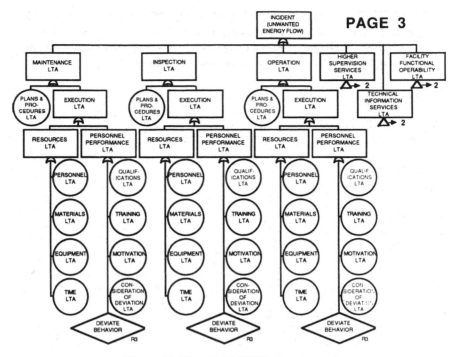

Figure 18-17   Mini-MORT chart, Page 3.

Plans and procedures: Were adequate maintenance plans and procedures in place? Did the plans and procedures match the operation? Were they current? Were they well written, technically correct, and understandable by maintenance personnel? Were they readily available to maintenance personnel? Did the plans cover both preventive and corrective maintenance requirements?

Execution: Were the maintenance plans and procedures followed? Evaluate the following factors:

Resources: Did the maintenance organization have the resources needed to follow the maintenance plans and procedures effectively? Evaluate the following factors:

Personnel: Were a sufficient number of people available to perform the required maintenance operations? Did they have the skills and knowledge required? Was the organizational structure adequate (for example, adequate number of supervisors, appropriate worker-supervisor ratio)?

Materials: Were adequate supplies, repair parts, and other necessary materials available to support the maintenance effort?

Equipment: Was adequate equipment provided to support the maintenance effort? Was this equipment readily available and well maintained? Was maintenance equipment appropriate for the work being done? Were provisions for changing or upgrading maintenance equipment consistent with other changes in the system?

Time: Was sufficient time programmed into operations to perform preventive maintenance services? Were corrective maintenance operations performed in a timely manner? Was sufficient time allowed to perform corrective maintenance without creating time pressure On the maintenance and/or the production operations?

Personnel performance: Did the maintenance workers and supervisors perform their assigned tasks properly? Factors to be evaluated include

Qualifications: Were all maintenance personnel qualified in terms of education, training, and/or experience to perform assigned tasks? If special licenses or certifications were required, did personnel have them?

Training: Had personnel received the required training? Was it current? Was it adequate? Was it documented properly? Was it performance-based training?

Motivation: Were personnel motivated to perform properly? Was good performance rewarded or acknowledged? Was substandard performance acknowledged? Were barriers to performance in place?

Consideration of deviation: Did the maintenance program make allowances for human errors? Was a two-person concept used for critical operations? Did supervision monitor difficult, tedious, or other error-prone tasks?

Deviate behavior: Did the maintenance program address the possibility of intentional wrongdoing (sabotage) by maintenance or other personnel? Were critical systems or operations double-checked and tested after maintenance? Were personnel who may have been having emotional, marital, financial, or chemical abuse problems monitored? Was counseling provided or available? Could a single act by a single individual have produced catastrophic results? Had management identified, evaluated, and accepted risks associated with deviate behavior?

The *inspection* topics are concerned with the existence of an adequate inspection program. Factors to be evaluated include

Plans and procedures: Were adequate inspection plans and procedures in place? Did the plans and procedures match the operation? Were they current? Were they well written, technically correct, and understandable by inspection personnel? Were they readily available to inspection personnel?

Execution: Were the inspection plans and procedures followed? Evaluate the following factors:

Resources: Did the inspection organization have the resources needed to follow inspection plans and procedures effectively? Evaluate the following factors:

Personnel: Were enough people available to perform the required inspection operations? Did they have the skills and knowledge required?

Materials: Were adequate supplies and other necessary materials available to support the inspection effort?

Equipment: Was adequate equipment provided to support the inspection effort? Was this equipment readily available and well maintained? Was equipment appropriate for the tests and inspections performed? Were provisions made for changing and/or upgrading inspection equipment consistent with other changes in the system?

Time: Was sufficient time programmed into operations to perform inspection services? Were inspections performed in a timely manner? Was sufficient time allowed to perform inspections without creating time pressure on the maintenance and/or production operations?

Personnel performance: Did the inspectors perform their assignd tasks properly? Factors to be evaluated include:

Qualifications: Were all inspectors qualified in terms of education, training, and/or experience to perform assigned tasks? If special licenses or certifications were required, did all inspectors have them?

Training: Had personnel received the required training? Was it current? Was it adequate? Was it documented properly? Was it performance-based training?

Motivation: Were personnel motivated to perform properly? Was good performance rewarded or acknowledged? Was substandard performance acknowledged? Did barriers to performance exist?

Consideration of deviation: Did the inspection program make allowances for human errors? Was a two-person concept used for critical operations? Did supervision monitor difficult, tedious, or other error-prone tasks?

Deviate behavior: Did the inspection program address the possibility of intentional wrongdoing (sabotage) by inspectors or other personnel? Were critical systems or operations double-checked and tested after maintenance? Were personnel who may have been having emotional, marital, financial, or chemical abuse problems monitored? Was counseling provided or available? Could a single act by a single individual have produced catastrophic results? Had management identified, analyzed, and accepted the risks associated with deviate behavior?

The *operation* topics consider whether an adequate program was in place. Factors to be evaluated include

Plans and procedures: Were adequate operational plans and procedures in place? Did the plans and procedures match the operation? Were they current? Were they well written, technically correct, and understandable by operations personnel? Were they readily available to operations personnel?

Execution: Were the operations plans and procedures followed? Evaluate the following factors:

Resources: Did the operating organization have the resources needed to follow the operations plans and procedures effectively? Evaluate the following factors:

Personnel: Were enough people available to perform the required tasks? Did they have the skills and knowledge required?

Materials: Were adequate supplies and other necessary materials available to support operations?

Equipment: Was adequate equipment provided to support the operations effort? Was this equipment readily available and well maintained? Was equipment appropriate for the tasks and operations performed? Were provisions made for changing or upgrading equipment consistent with other changes in the system?

Time: Was sufficient time programmed into operations to perform inspection and maintenance services? Was sufficient time allowed to perform inspections and maintenance without creating time pressure on the production operations?

Personnel performance: Did the operators perform their assigned tasks properly? Factors to be evaluated include

Qualifications: Were all operators qualified in terms of education, training, and/or experience to perform assigned tasks? If special licenses or certifications were required, did operators have them?

Training: Had personnel received the required training? Was it current? Was it adequate? Was it documented properly? Was it performance-based training?

Motivation: Were personnel motivated to perform properly? Was good performance rewarded or acknowledged? Was substandard performance acknowledged? Did barriers to performance exist?

Consideration of deviation: Did the operations program make allowances for human errors? Was a two-person concept used for critical operations? Did supervision monitor difficult, tedious, or other error-prone tasks?

Deviate behavior: Did the operations program address the possibility of intentional wrongdoing (sabotage) by operators or other personnel? Were critical systems or operations double-checked and tested after maintenance? Were personnel who may have been having emotional, marital, financial, or chemical abuse problems monitored? Was counseling provided or available? Could a single act by a single individual have produced catastrophic results? If the system was vulnerable to deviate behavior, had management identified, evaluated, and accepted this risk?

The next major branch of the Mini-MORT chart to be considered addresses the barriers to the unwanted energy flow (Fig. 18-18). Even though physical barriers are of primary importance, administrative and procedural barriers are also included. Barriers can be placed on the energy source (such as insulation on electrical wires), between the energy source and potential targets (fence around electrical substation), or on persons or objects that may be targets (personal protective gear worn by electrical workers). Many operations, of course, use a combination of barriers. The analyst identifies and lists all the barriers that were (or should have been) involved in the accident and then evaluates the failure of each to prevent the unwanted energy flow by evaluating the following items:

Design: Was the initial design of the barrier sufficient to prevent the unwanted energy flow? Was it configured properly? Were the proper materials used? Were the materials the proper strength? Did the design have human factors deficiencies? Were the drawings and specifications prepared properly? Were the design and design documents reviewed properly? Were the safety and user communities included in the design and review process?

Fabrication: Was the barrier built and/or manufactured as designed? Did it meet specifications? Was quality control adequate? Was testing adequate?

Installation: Was the barrier installed properly? Was it installed (or worn or used) in accordance with the design or with manufacturer's specifications? Was it installed in the correct location and position? Were installation procedures provided? Were they followed? Was the installation performed by qualified personnel?

Maintenance: Was the barrier maintained properly? Was it inspected and tested as required? Was the barrier disabled or damaged by main-

PAGE 4

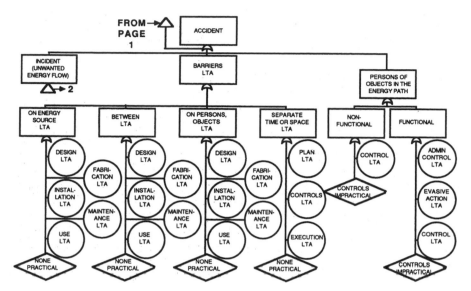

Figure 18-18   Mini-MORT chart, Page 4.

tenance operations? Were maintenance instructions provided? Were they followed? Was maintenance performed by qualified personnel?

Use: Were barriers used? Were they provided? Were personnel instructed in the requirement to use the barriers? Were they trained to use the barriers? Did supervision enforce use of barriers? Did operating instructions, job safety analyses, or other relevant procedures require the use of barriers? Were the barriers difficult to use or did they cause discomfort and/or interfere with operations or task performance? Were they available at the appropriate time and place?

None practical: If barriers were not practical for the energy source used, had management identified, evaluated, and accepted the risk?

Another type of barrier is *separate time or space.* The most practical barrier may be simply vacating the area, which works well with the nuclear weapons testing program, gunnery ranges, demolition work, and other places where a large quantity of relatively uncontrolled energy may be released in a relatively short time. To evaluate this type of barrier, consider the following:

Plan: Was an adequate plan in place for ensuring that the area has been vacated? Did the plan provide for orderly, controlled evacuation? Did the plan include methods of accounting for evacuated personnel and

property? Did the plan have adequate provisions for ensuring that unauthorized persons or objects could not enter the danger area or enter
at a hazardous time?

Controls: Were the controls provided as specified in the plan? Were they
adequate? Did they provide timely protection, and did they provide verification that the appropriate areas were evacuated at the appropriate
times?

Execution: Did everyone do what they were supposed to do? Did control
systems work properly? Was the barrier effective?

None practical: Did management identify, evaluate, and accept the risks
associated with the operation?

The final "accident" branch to be considered concerns persons or objects
in the energy path. They are the people who were injured and/or the objects
that were damaged and represent the losses from the accident. The purpose
of this branch is to determine if the losses could have been prevented by
moving the targets out of the path either before or after the energy release.
Items to be evaluated include

Nonfunctional: If the persons or objects in the energy path were not there
as a necessary part of the system, they are nonfunctional. This class generally includes visitors, passers-by, the great American public, intruders,
and perhaps even organizational members with no requirement to be at
the accident location when the accident occurred. Factors to be evaluated include

  Control: Were adequate controls in place to ensure that persons or
  objects not required for the operation were not exposed to potential
  hazards?

  Controls impractical: Were the risks associated with performing the operation without controls to protect persons or objects in the event of an
  unwanted release of energy identified, analyzed, and accepted by the
  proper level of management?

Functional: Functional persons or objects are those that were required to
be at the accident location at the time of the accident as a necessary part
of the system. The individuals and/or objects were at their proper duty
station performing assigned tasks or working as a necessary part of the
system. Evaluate the following:

  Administrative controls: Were steps taken to organize the work and the
  workplace to minimize exposures?

  Evasive action: Were plans made and steps taken to remove persons or
  objects from the energy path? Were these actions successful?

  Controls: Were control measures to warn targets of impending danger
  adequate?

Controls Impractical: Were the risks associated with performing the operation without controls to protect persons or objects in the event of an unwanted release of energy identified, analyzed, and accepted by the proper level of management?

The root causes of most accidents are found during an analysis of the management system factors (Fig. 18-19). This major branch of the chart addresses the why issues. Specific factors and branches to be analyzed include

Policy: Did the organization have a clear, well-written, widely distributed policy statement outlining management's commitment to safety and the corporate safety philosophy? Are the rights and responsibilities of employees included? Was the policy statement current and signed by the appropriate manager? Were well-defined safety goals and objectives included?

PAGE 5

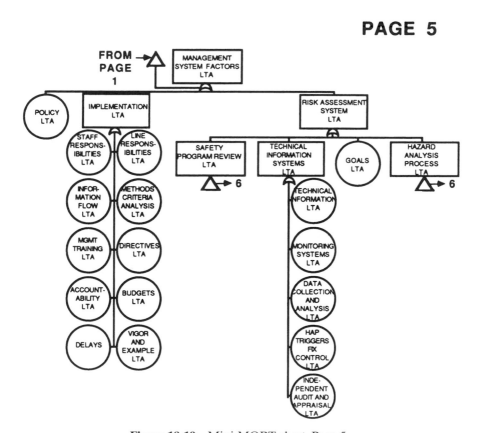

Figure 18-19 Mini-MORT chart, Page 5.

Implementation: Did the organization function as advertised and take the appropriate steps to meet the goals and objectives contained in the safety policy? Evaluate the following:

Staff responsibilities: Were staff responsibilities clearly defined in terms of the safety program and emergency responses? Did recruiting and hiring practices ensure that the organization employed qualified personnel? Was a medical screening and medical profile program in place that verified that individuals were physically able to perform assigned work? Did the program include periodic physical examinations? Did procurement policies support the safety effort by ensuring that only approved tools and equipment were provided? Were safety and user input and review part of the procurement process? Was safety review part of subcontractor selection? Were subcontractor safety requirements provided in requests for proposals, prebid and preconstruction conferences, and contracts, as appropriate? Were the requirements adequate? Did the safety (including environmental, health, safety, and fire protection) staff have the appropriate responsibilities for providing support to the line organization? Was adequate quality assurance provided? Were responsibilities for providing other necessary staff support functions clearly defined?

Line responsibility: Were line responsibilities clearly defined throughout the organization? Did the organization clearly state that safety is a line responsibility and that supervision and management at all levels were responsible for the health and safety of the people and operations they supervise? Were the safety rights and responsibilities of individual employees included?

Information flow: Were adequate provisions made for providing a free flow of information both vertically and horizontally within the organization? Did a formal program of staff meetings and safety meetings provide rapid verbal distribution of significant safety and operational information? Was it adequate? Was the distribution of verbal directives documented? Was written information distributed effectively? Was it verifiable? Was it adequate? Were alternate channels of communication available for reporting safety concerns? Were they adequate?

Methods, criteria, and analysis: What management methods and techniques were employed to ensure that safety programs and policies were implemented (management/safety by objectives, for example)? Were these methods adequate? What criteria were used to evaluate the effectiveness of implementation? What types of qualitative or quantitative analyses were done to determine the effectiveness of safety program implementation? Were they adequate?

Management training: What safety education and training did managers have? Was an established safety training program for management in place? Was it adequate?

Directives: Had the appropriate directives or other implementing documents been issued to support the safety program and implement the safety policy? Were they adequate? Were they distributed to the proper levels of the organization?

Accountability: Were managers at all levels held accountable for implementation of the safety program and for meeting their safety responsibilities? Were management performance appraisals, promotions, bonuses, and raises based, in part, on safety performance? Was this basis known? Did safety performance affect budgets (were accidents charged back to operations)?

Budgets: Was the safety effort adequately funded? Did line and staff organizations have the money necessary to perform assigned tasks properly without taking shortcuts or using unsafe tools, equipment, or facilities? Were budgets adequate to ensure that line and staff organizations were able to hire, train, and retain adequate numbers of qualified personnel? Were projects funded adequately to ensure that necessary safety analyses and system safety efforts could be performed and that subcontractors were well qualified?

Delays: Were safety resources provided in a timely manner? Were safety problems prioritized and acted on without undue delays?

Vigor and example: Did management set high personal safety standards? Was management's participation in safety meetings, safety committees, safety award presentations, and other safety functions highly visible? Did management participate in work site safety inspections?

Risk assessment system: For management to make enlightened safety and risk management decisions, accurate, timely risk assessment information must be provided. Risk assessment system factors to be evaluated include

Goals: Were the goals and objectives of the risk assessment effort clearly stated and communicated? Are they consistent with organizational goals and the safety policy? Are they adequate?

Technical information systems: Timely, accurate, relevant technical information is essential to any risk assessment effort. The same factors should be evaluated here that were evaluated under "technical information systems" on the specific control factors side of the chart, but this time the evaluation is from the perspective of technical information as it applies more broadly to the overall risk assessment system. On the specific control factors side of the chart, the emphasis was on the adequacy of technical information systems as they applied to the specific unwanted energy flow involved in the accident.

Technical information: Was adequate, accurate information about energy sources, barriers, and targets known? Was it communicated effectively to all of the individuals needing the information? Were all available information sources used (for example, external data banks, consultants, studies, and current research)?

Monitoring systems: Were the monitoring systems in use appropriate and sufficient? Were the monitoring systems operating properly? Was the information from the monitoring effort valid, useful, and communicated effectively? Were a mix of monitoring techniques employed?

Data collection and analysis: Was available information collected, distributed, and used properly? Was information collected, analyzed, and communicated in a timely manner? Was the analysis accurate and the results meaningful? Were data collected from different sources correlated and cross-checked?

Hap Triggers and fix control: Was an adequate plan in place to initiate the hazard analysis process (HAP) and fix controls based on input from the monitoring system or from other sources of technical information?

Independent audit and appraisal: Were provisions made for objective "outside eyes" to perform audits and appraisals? Had independent audits and appraisals been performed? Recently? Were they adequate? Had the results been communicated and used properly?

The *safety program review* section is concerned with whether reviews and other support of the risk assessment effort were adequately provided (Fig. 18-20). Factors to be evaluated include

Higher supervision services: On the specific control factors side of the chart, higher supervision services are evaluated to determine the adequacy of operational support provided. Here, the same factors are evaluated to determine the adequacy of support to the risk assessment effort.

Goals: Were the job goals and safety goals set by higher supervision realistic, obtainable but challenging, and consistent with organizational goals?

Information exchange: Was information about job tasks, safety requirements, hazards, rights and responsibilities, and other relevant information passed accurately and in a timely manner between direct supervision and management in both directions? Did middle levels of supervision shortstop, filter, or inappropriately alter communications? Did middle supervision interpret or supplement information as necessary?

Research and fact-finding: Did middle supervision search out and provide the information necessary to support the review process?

Standards and directives: Did middle supervision maintain the appropriate standards and directives? Were they kept up-to-date and accessible?

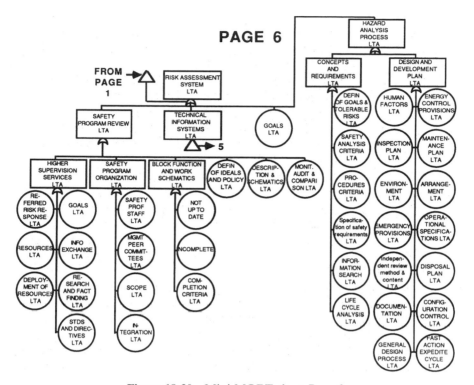

Figure 18-20 Mini-MORT chart, Page 6.

Referred risk response: Were supervisors responsive when hazards were reported to them? Did problems, concerns, or requests for assistance in analyzing or reviewing risks receive appropriate responses?

Resources: Did middle supervision ensure that the necessary personnel, tools, equipment, procedures, and time were available to do prescribed tasks?

Deployment of resources: Did middle supervision ensure that available resources were distributed fairly and consistent with job needs?

Safety program organization? Did the safety program contain all the necessary elements, and were they organized efficiently and effectively? Was the program primarily proactive? Evaluate the following:

Safety professional staff: Did the safety staff have adequate professional qualifications? Did safety professionals have professional registration (certified safety professionals or professional engineers)? Did safety professionals have the educational backgrounds and credentials necessary for technical competence and credibility in the organization? Did they have an appropriate mix of safety, engineering, human factors, and management skills and knowledge? Did the safety staff have adequate knowledge and experience with the organization and operations? Did the safety staff actively participate in professional development activities and professional organizations?

Management peer committees: Did the safety program include safety committees with management, supervision, safety, and worker representation? Were the committees well supported, active, and effective?

Scope: Did the safety program include a comprehensive safety education and training program? Did the safety program include provisions for ensuring compliance with applicable codes, standards, and regulations? Did the safety program include a comprehensive inspection, audit, and appraisal program? Did the safety program include a system safety effort with provisions for safety and user participation in design, procurement, and review efforts? Did the safety program include provisions for adequately meeting administrative requirements? Did the safety program include an effective accident investigation and analysis program? Did the program provide adequate provisions for monitoring and evaluating safety performance?

Integration: Was the safety program adequately integrated into the overall organization and operations? Did it have the clout necessary to be effective?

Block function and work schematics: Were charts, drawings, diagrams, and other graphic documents adequate?

Not up to date: Were drawings and other graphic documents current?

Incomplete: Were "as built" documents available and accurate?

Completion criteria: Were the requirements and criteria for completing and updating graphic documents adequate?

Definition of ideals and policy: Were safety program review goals and objectives clearly defined and adequate?

Description and schematics: Were the safety program and safety program review process adequately described to include review matrices and organizational charts?

Monitoring audit, and comparison: Were adequate provisions made to evaluate the safety program and the safety program review effort? Were meaningful comparisons with similar organizations or operations included?

The *hazard analysis process* considers whether the tools, techniques, plans, and requirements for performing hazard analysis are adequate. Factors to be evaluated include

Concepts and requirements: Was the scope of the hazard analysis process adequate? Evaluate the following:

Definition of goals and tolerable risks: Were risk acceptability criteria clearly defined? Were they consistent with safety requirements, yet realistic and obtainable?

Safety analysis criteria: Was the protocol for selecting projects, items of hardware, or operations for detailed analysis clearly defined and adequate? Were appropriate hazard analysis techniques recommended or specified?

Procedures criteria: Were the requirements and criteria for developing operating, maintenance, and other procedures clearly defined and adequate?

Specification of safety requirements: Were applicable codes, standards, and regulations called out? Were additional safety requirements clearly stated? Were they adequate?

Information search: Was relevant existing information researched?

Life cycle analysis: Did the hazard analysis process include the entire life cycle?

Design and development plan: Was the design and development effort adequate? Evaluate the following:

Human factors: Were stereotypes violated during the design process? Were ergonomic principles properly applied? Were human-machine interfaces adequately evaluated? Was adequate misuse analysis performed?

Energy control provisions: Were the safest forms of energy used, consistent with operational requirements? Were adequate barriers provided? Was the safety precedence sequence properly applied? Were energy flows carefully considered and routed away from vulnerable targets where possible?

Inspection plan: Were inspection requirements considered during design and adequate provisions made for safely inspecting critical components?

Maintenance plan: Was maintainability adequately considered? Could the system be maintained without compromising the safety of maintenance personnel? To what degree could the system tolerate poor maintenance without catastrophic failure?

Environment: Did the design and development process accurately and adequately consider the environmental conditions in which the system would operate?

Arrangement: Was the system arranged in such a way that operations, housekeeping, storage, material handling, and other activities can be done with minimum exposure to hazards? Were high energy sources kept away from vulnerable targets? Were displays and signs designed and arranged to expedite accurate monitoring and interpretation?

Emergency provisions: Were systems designed to fail safe? Were life safety codes met? Were high human error rates in emergency situations considered?

Operational specifications: Were performance specifications and operating parameters clearly defined? Were they reasonable and adequate?

Independent review of method and content: Were review points adequately defined in the design and development plan? Were review methods or techniques specified? Were review responsibilities and review criteria adequately defined? Was adequate time provided for review?

Disposal plan: Did the design and development plan consider potential problems associated with decommissioning, demolition, or disposal of the system at the end of the life cycle?

Documentation: Were design documents carefully controlled to ensure only the correct (current) documents were distributed? Was documentation accurate and adequate?

Configuration control: Were changes managed and documented properly? Were appropriate reviews and counterchanges made during the design and development process? Were "field changes" or other unauthorized modifications controlled?

General design process: Was the design process managed in a professional manner? Were competent, qualified personnel involved in design and review? Were appropriate safety analyses performed?

Fast action expedite cycle: Were provisions made in the design and development plan to respond rapidly to priority requirements without compromising project integrity or safety?

These questions were phrased in the past tense to indicate that the analysis should focus on conditions as they were at the time of the accident. If the analysis is being performed as a preventive exercise, the questions are obviously phrased in the present tense to evaluate current conditions.

After completely evaluating the chart, cross-check results with the event and causal factors chart and change analysis. Reconcile any differences, use all three tools to aid in writing the formal accident report, and be careful to provide findings, conclusions, and recommendations (backed up with evidence) for each factor deemed to be less than adequate.

REVIEW QUESTIONS

1. What is the greatest advantage of MORT analysis?
2. What is the greatest limitation?
3. Why was the mini-MORT chart developed?

4. When should MORT analysis be initiated for an accident investigation?
5. What are assumed risks?

REFERENCES

Johnson, William G. 1973. *MORT, the MANAGEMENT OVERSIGHT AND RISK TREE.* Washington, DC: U.S. Atomic Energy Commission.

Johnson, William G. 1980. *MORT Safety Assurance Systems.* New York: Marcel Dekker.

Knox, N. W. and Eicher, R. W. 1983. *MORT User's Manual.* SSDC-4 (Revision 2) Idaho Falls, ID: U.S. Department of Energy.

CHAPTER 19

Event and Causal Factors Charts

The basic concept from which event and causal factors charts were developed can probably be traced back to Ludwig Benner and others at the National Transportation Safety Board. Benner developed a very similar technique called multilinear event sequencing (MES) and more recently sequentially timed events plotting (STEP). Event and causal factors charts were part of the overall MORT approach to system safety developed by W. G. Johnson for the Atomic Energy Commission in the early 1970s and further developed and taught by the Department of Energy's System Safety Development Center (SSDC). The use of the event and causal factors chart is sometimes referred to as causal factors analysis.

Event and causal factors charts are graphic representations that basically produce a picture of an accident—both the sequence of events that led to the accident and the conditions that were causal factors. This tool works very well in conjunction with PET or MORT analysis and is used widely in the Department of Energy.

Unlike PET, MORT, and mini-MORT charts, which are working papers to be filed with backup documents or discarded after use, the event and causal factor chart is an important descriptive tool. An "executive summary" version of the larger, more detailed chart developed during the investigation is included in the formal accident report as an illustration. The event and causal factors chart aids initially in collecting and organizing evidence and in developing an understanding of the accident sequence and causes. It is extremely valuable as an aid in writing the report and makes an excellent briefing tool.

PURPOSE

The purpose of the event and causal factors chart is to provide a systematic accident analysis tool to aid in collecting, organizing, and depicting accident

System Safety for the 21st Century: The Updated and Revised Edition of System Safety 2000, by Richard A. Stephans
ISBN 0-471-44454-5   Copyright © 2004 John Wiley & Sons, Inc.

information; validating information from other analysis techniques; writing and illustrating the accident report; and briefing.

INPUT REQUIREMENTS

Detailed information about the accident sequence and the specific and systemic conditions surrounding the accident are required.

GENERAL APPROACH

Like other accident analysis techniques, the event and causal factors chart should be initiated as soon as the investigation begins and developed as evidence is collected. A detailed chart is developed during the investigation; it is then reduced to an executive summary containing only key events and causal factors to serve as a report illustration.

INSTRUCTIONS

Several self-sticking notepads (two different colors and/or sizes are recommended, one for events and one for conditions) or index cards and masking or cellophane tape are required, as are a bare wall, table, and bulletin board or blackboard, preferably in a private working location that will be undisturbed and used exclusively by the investigator(s) until the work is complete.
   The symbols used are

1. Rectangles for events
2. Circles for accident events
3. Ovals for conditions
4. Arrows to connect events and conditions
5. Dotted lines for presumptive events and conditions

Ideally, events and conditions can be supported by hard evidence. Occasionally, however, an event or condition can be based only on logic and the best available evidence. Such data are called *presumptive* events or conditions. They may be necessary to complete an accident sequence or to explain the most likely causes of the accident.
   Start the event and causal factors chart as early as practical. From the available information, make an event note or card for each event that occurred during the accident sequence and a condition card for each relevant condition that is identified. Organize and reorganize the cards as more information becomes available.

The primary chain of events that led to the accident should be organized chronologically from left to right across the middle of the bulletin board or other surface. This time line forms the primary sequence line. Any secondary events (other relevant actions going on during the same time as the primary events) are also sequenced from left to right and are depicted either just above or just below the primary event line (Fig. 19-1). Specific conditions that can be linked to any particular event are indicated in an oval above or below that event and linked by arrows to the events they describe. Similarly, event symbols are connected with arrows. Systemic conditions—that is, conditions that may have been preexisting and that continued throughout the accident sequence—are indicated farthest from the primary event line and may not be connected to any particular event.

The rules for developing an event and causal factors chart include

1. Dates and/or times are indicated just below primary accident sequence events.
2. Events are actual occurrences described by a noun and an action verb.
3. Each event symbol contains only one discrete event.
4. Events and conditions are quantified if possible.

The event and causal factors chart is developed and expanded as new evidence and information become available. It should also be developed in conjunction with other analyses (change analysis and PET or MORT analysis), and information should be cross-checked (Fig. 19-2). Conditions that are causal factors on the event and causal factors chart should be identified as less than adequate areas on the PET, MORT, or mini-MORT chart.

The completed chart, along with the other analytical tools, is used in writing the accident report. The primary event line provides the chronology of events.

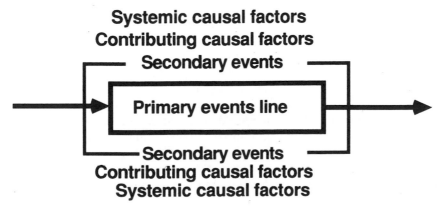

Figure 19-1 Event and causal factors relationships.

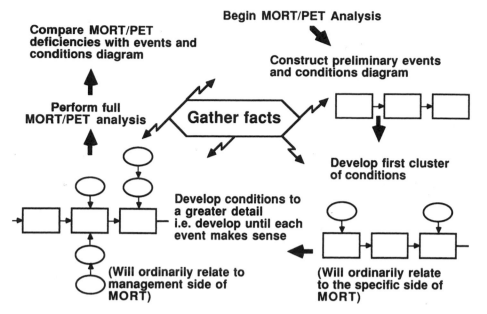

Figure 19-2 Relationships between PET or MORT and events and conditions. The event and causal factors chart and the PET, MORT, and/or mini-MORT chart are initiated as soon as the investigation begins and are developed together as evidence is collected.

The conditions are causal factors and translate into findings and conclusions in the report.

Event and causal factors charts can be relatively simple and straightforward. The complexity of the chart is determined by the complexity of the accident and the desired level of detail.

Other than determining the appropriate level of detail, two other key decisions must be made. The starting point for the event sequence must be determined, and the appropriate stopping point or final event must be found.

The appropriate starting point is sometimes difficult to determine. A general rule of thumb is to start with the earliest event that led directly to the accident. This event may have occurred on the day of the accident or days, weeks, or even months earlier. Obviously, this decision can be very subjective.

Providing specific guidance on the stopping point is a little easier. Generally, the stopping point is not the accident event but rather the postaccident event that indicates that the situation is stabilized and that direct losses are no longer occurring. For example, the logical stopping point for most fires is the event indicating that the fire is out. For an injury accident, post-accident events normally continue until the injured have been transported to a medical facility.

In most cases, including long-term events or losses such as litigation, regulatory fines, or long-term medical expenses in the chart is not practical.

Sample Event and Causal Factors Chart

Figure 19-3 indicates the manner in which a chart would be arranged. It shows the approximate number of events and conditions that would be expected on an executive summary type of chart for a relatively minor accident.

It outlines the sequence of events and the conditions that were causal factors in the overturning of a small hydraulic crane. The crane was set up properly at 7:30 A.M. on the day of the accident in a position to off-load two truckloads of materials in a layout yard. The area was level and compacted and provided adequate ground support for the crane.

The first truck arrived at approximately 7:42 A.M. Patches of fog in low-lying areas of the road had led the driver to turn on his lights.

After positioning the truck in the area adjacent to the crane, the driver turned off the engine and walked to the yard office to use the rest room and call his girlfriend. He inadvertently left the truck lights on.

The crane operator off-loaded the truck, and at about 8:40 A.M. the truck driver returned and tried to start the truck. However, the lights had run down the battery to the point where it would not start. About this time, a second truck arrived. Unable to pull into the desired off-loading location, the driver parked on the opposite side of the crane. The crane could reach the truck without relocating, but the load radius was fifty-four feet rather than twenty-seven feet. The crane operator started to off-load the second truck at about 8:52 A.M. When he started to swing the load, however, the crane overturned as the load cleared the truck (approximately 8:55 A.M.). The accident caused no injuries and no damage to either of the trucks or to the load, but the crane boom, cab, and one outrigger were damaged (repair estimate of $16,000).

The site was secured, and supervision was notified as outlined in company procedures by 9:10 A.M.

The crane was properly certified, but the investigation revealed that the load chart was faded, torn, and generally unreadable. The operator had attempted to make the lift without consulting the load chart and had exceeded the radius for the load and crane configuration; thus, the crane overturned.

The working chart used during the investigation would contain much more information and detail than this simplified chart. Events and conditions would address the truck maintenance program, the qualifications and conditions of the drivers and crane operator, and obviously more detail about the models, makes, and conditions of the trucks and crane, details about the load (weights and rigging, for example), and more background information about the yard and the operations.

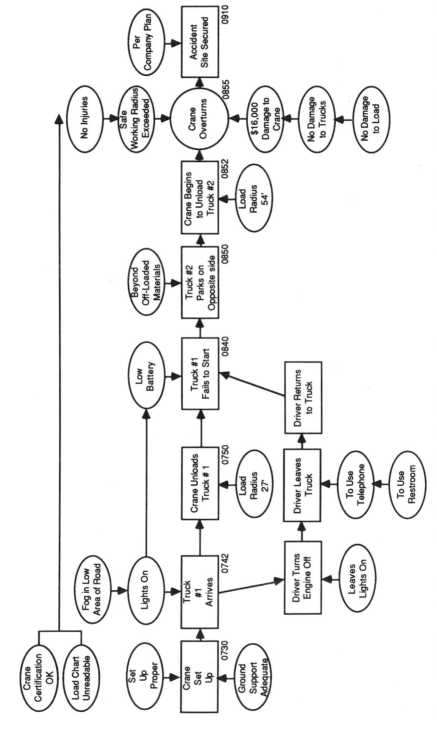

Figure 19-3   Example of an event and causal factors chart. This executive summary version of the chart would be included in the formal accident report and used as a briefing tool.

REVIEW QUESTIONS

1. What are the relationships among event and causal factors charts, MES, and STEP?
2. At what point in the investigation should an event and causal factors chart be initiated?
3. What event symbol is used to indicate the actual accident event?
4. At what point should the event and causal factors chart normally be stopped? At what event?
5. Name four uses for the event and causal factors chart.

REFERENCES

Johnson, William G. 1973. *MORT, the Management Oversight and Risk Tree.* Washington, DC: U.S. Atomic Energy Commission.

Johnson, William G. 1980. *MORT Safety Assurance Systems.* New York: Marcel Dekker.

CHAPTER 20

Other Analytical Techniques

In addition to the analysis techniques discussed in Chapters 13 through 19, many other approaches, methods, and techniques are used to perform system safety analyses.

Some of these different techniques actually represent new or unique approaches or methods. Others are variations of different names for common techniques.

This chapter describes some of these techniques that may be of value in specific system safety efforts and those that the system safety practitioner is likely to encounter in system safety literature.

SOFTWARE HAZARD ANALYSIS

Software hazard analysis is extremely important and will be of growing importance in the future. The software hazard analysis effort should parallel the system safety program for system hardware, be a life cycle effort, and use a combination of methods and approaches.

The analysis effort for software should address both software requirements and the software codes and programs. Both operating system and applications software should be included.

Unfortunately, the state of the art in software hazard analysis appears to be woefully lagging. Even though traditional hazard analysis techniques like fault tree analysis and tailored versions of operating hazard analysis may be applied to the evaluation of software, validated, specific methods of software hazard analysis appear lacking.

Ironically, the most promising methods of analyzing software are software packages specifically designed to perform software hazard analyses and diagnostic features built into critical software programs.

System Safety for the 21st Century: The Updated and Revised Edition of System Safety 2000, by Richard A. Stephans
ISBN 0-471-44454-5   Copyright © 2004 John Wiley & Sons, Inc.

COMMON CAUSE FAILURE ANALYSIS

Common cause failure analysis (CCFA) or common cause analysis is used primarily to evaluate multiple failures that may be caused by a single event or causal factor common to or shared by multiple components. It is especially important in evaluating the true reliability produced by redundant systems or components.

Common causes of failure can be a common or shared environment or location, common (same) manufacturer, common maintenance or operations personnel and/or procedures, common power and/or fuel supply or circuits, common control signals or circuits, common design specifications, and a common service life.

If redundant elements are relied upon to provide the required level of safety and reliability in critical systems, these elements must not be subject to failure caused by a common single event or common causal factors.

During the late 1980s two air tragedies resulted from very similar common cause failures. In both cases (a Japan Air Lines Boeing 747 and a United Airlines DC-10), hydraulic control was lost when the redundant hydraulic systems (seven in the 747 and three in the DC-10) were disabled by loss of hydraulic fluid when the hydraulic lines in the rudder of the aircraft were severed. Only one hydraulic system in either aircraft would have been sufficient to maintain control of the aircraft. The only location on either aircraft where all the hydraulic lines were in close proximity was in the rudder. Even though the trigger event was different for each aircraft (aft pressure bulkhead failure and sudden depressurization in the 747 and an engine explosion in the DC-10), both aircraft crashed due to common cause failure of redundant hydraulic control systems.

A general approach to common cause failure analysis is to identify critical systems or components and then use energy trace and barrier analysis (ETBA) to evaluate vulnerability to common environmental hazards, unwanted energy flows, and barrier failures (see Chapter 13).

The project evaluation tree (PET) can be used to evaluate and analyze common operating and maintenance procedures and personnel, as well as common acquisition, construction, installation, testing, and manufacturing processes (see Chapter 16).

SNEAK CIRCUIT ANALYSIS

A sneak circuit or path is an unintended energy route, which can allow an undesired function to occur, prevent desired functions from occurring, or adversely affect the timing of functions. Sneak circuit analysis or sneak analysis is performed to identify ways in which built-in design characteristics can either allow an undesired function to occur or prevent desired functions from occurring. Even though most sneak circuits and most sneak circuit analysis efforts involve electrical or electronic circuitry or devices, sneak circuits can also occur in hydraulic, pneumatic, and mechanical systems and in software.

An important feature of sneak circuit analysis is that the sneak paths being investigated are not the result of a component failure. They are rather the result of the circuit design. The sneak paths may show up only on rare occasions when the switches (or valves) in the circuit are in a unique configuration.

Sneak circuit analysis is usually inductive and can be very difficult to perform without the software to aid in producing network trees and other graphics. Much of this software is proprietary, and a large portion of the sneak circuit analysis work done to date has been accomplished by large aerospace and weapons contractors (notably Boeing and General Dynamics).

The most common approach to sneak circuit analysis involves visual clues found by comparing circuits with the five basic topographs shown in Figure 20-1. Nearly every circuit can be broken down into combinations of these topographs.

Sneak conditions to be sought include

1. Sneak paths that can cause current (or another type of energy) to flow along an unexpected route
2. Incompatible hardware or logic sequences that can cause unwanted or inappropriate system responses (sneak timing)
3. Sneak indications that can cause misleading, ambiguous, or false displays
4. Sneak labels that provide confusing or incorrect nomenclature or instructions on controls and that thus promote operator errors.

Successful sneak circuit analysis requires a skilled analyst and great care. Unfortunately, sneak circuit analysis is frequently performed with

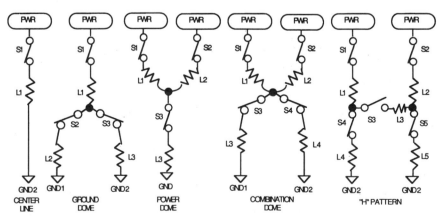

Figure 20-1 Sneak circuit topographs. (Source: NASA: NSTS 22254, Methodology for Conduct of NSTS Hazard Analyses.)

more accuracy as an accident analysis tool than as an accident prevention tool.

EXTREME VALUE PROJECTION

Extreme value projection is a risk projection technique that is quick and easy to use and yet is a true projection technique in that it can predict or provide information about losses that are more severe than any losses that have occurred to date. For example, it can project the probability and/or return period for a $500,000 loss even though the most serious loss to date has not exceeded $350,000.

Therefore, extreme value projection may be useful in calculation probabilities to be used for risk assessment codes or total risk exposure codes based on actuarial data. Extreme value projection can also be useful in communicating hazard information to management and may aid in drawing management attention to a problem before a serious accident occurs.

For accident analysis, extreme value projection is important because it can help to determine whether the specific accident being investigated was typical of the system or a true outlier. If the accident was typical of the system, the investigation concentrates on system shortcomings and recommendations to prevent recurrence and includes systemic changes. If the accident was a true outlier and not typical of the way the organization operates, the investigation concentrates on the specific causes of this particular accident and uses change analysis to determine what made the actions and events surrounding the accident atypical. Recommendations should concentrate on change controls and not necessarily change major elements of the safety program or other parts of the system that may be performing well.

Extreme value projection requires a special graph paper produced by the TEAM (Technical and Engineering Aids for Management Company, Box 25, Tamworth, New Hampshire 03886; 603-323-8843). The paper is available with either log/normal scales (for most applications) or log/log scales for situations in which loss rates have increased very rapidly (Fig. 20-2).

As the name implies, the technique requires extreme or worst-case loss data, which are generally easy to obtain.

The special-purpose paper (TEAM #111 for log/normal and TEAM #112 for log/log) already has most of the scales preprinted. The x-axis at the bottom of the paper contains the cumulative probability scale; the x-axis at the top of the paper contains the return period scale. The y-axis is used for plotting the magnitude of the particular type of losses being investigated and is an analyst-defined scale.

Extreme value projection can be done for any type of loss that can be quantified. The first step in the process is to select the type of loss to study (property damage, lost time, fire damage, or vehicle damage, for example) and the time span for the study or analysis (such as last year, last two years, or last ten

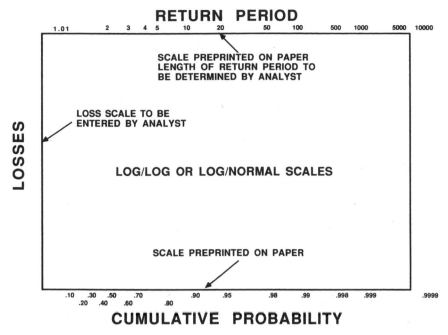

Figure 20-2 Scales used on the special graph paper required for extreme value projection.

years). Because change can invalidate any risk projection technique, the operation being studied should be free of significant changes for the time span of the study. This absence of change generally means that the location, process, method of operation, product, key management personnel, procedures, and policies must remain relatively constant. If significant changes have occurred during the time span, the data may not plot as a straight line and will probably contain a distinct dogleg.

The second step is to select the length of the return period. The *return period* is a slice of the total time span of the analysis. The time span to be analyzed should have at least five and no more than twenty return periods. Therefore, a return period of a month would be a logical choice for a study to cover a year, and a return period of a year would be appropriate for a ten-year study.

Step three is to collect the worst-case data for the particular category of loss being investigated for each return period. For a ten-year study of fire damage, for example, identify the worst loss (in dollars) caused by fire during each year of the ten-year period (assuming the return period selected is one year).

The fourth step is to rank order the data (the worst cases from each return period) from low to high.

The fifth step is to calculate the cumulative probability for each item by dividing the item number from the rank-ordered list by the total number of items (return periods). For instance, if the study covers a ten-year period and the return period is one year, the cumulative probability for the first item (the lowest value of loss during the ten-year period) would be

$$\frac{1 \text{ (the first item)}}{10 \text{ (the number of items)} + 1} = \frac{1}{11}$$

Unless a compelling reason to do otherwise exists, use nine items (return periods) for each study (nine months, nine quarters, or nine years, for example). This number of data points is appropriate, and, as the number of items 9 plus 1 is 10, the cumulative probability for each item can be calculated simply by shifting the decimal point (cumulative probability for item 1 = 0.1, item 2 = 0.2, item 3 = 0.3, and so on).

The sixth step is to select the scale for the loss axis. The scale should be selected so that the highest value of the losses during the time span of the study plots between half the way and three quarters of the way up the y-axis. This value provides an opportunity to project information about losses of a greater magnitude than those experienced during the time span of the study and thus perform true risk projection.

The seventh step is to plot the points by using the cumulative probability for the x-axis and the value of the loss for the y-axis (see the example).

The next step is to draw a best fit straight line through the points so that roughly as many points fall above the line as below. The line does not actually have to pass through any given point.

The final step is to evaluate and use the results from the effort. If the data points do not approximate a straight line, the analysis is not valid for the time span selected (significant changes occurred) and/or the appropriate paper was not selected (a curved line on log and normal paper should plot as a straight line on log and log paper).

If the extreme value projection is performed as part of an accident analysis, the technique is valid and fits the data if all points approximate a straight line *or* if all of the points but one (the one representing the accident) fall on a straight line. If all points, including the point representing the accident, fall on a straight line, then the accident was typical of the operation and could have been predicted. In order to prevent recurrence of similar accidents, systemic changes may be required, and the root causes of the accident probably are upstream in the management system.

By contrast, if all points but the point representing the accident approximate a straight line and the point representing the accident falls significantly above the line, then the accident was a true outlier and was not typical of the system. Recurrence may be prevented by identifying the reasons the accident situation was different from normal operations by using change analysis. Significant changes in the system may not be appropriate.

An example illustrates how extreme value projection can be used as a risk projection tool. The purpose of the analysis is to determine how often a manu-

TABLE 20-1. Example

Quarter	Worst Loss in Dollars	Rank Ordered Losses	CP
1 1991	42,000	1. 2,000	.1
4 1990	18,000	2. 4,000	.2
3 1990	30,000	3. 10,000	.3
2 1990	2,000	4. 18,000	.4
1 1990	67,000	5. 28,000	.5
4 1989	10,000	6. 30,000	.6
3 1989	4,000	7. 42,000	.7
2 1989	59,000	8. 56,000	.8
1 1989	28,000	9. 67,000	.9

facturing operation can expect to suffer a $90,000 (upper limit of insurance coverage) property damage accident. The manufacturing operation is relatively new and has never had a single property damage loss greater than $67,000. Because the operation is only about three years old, the study covers the last nine quarters for which data are available (see Table 20-1).

Figure 20-3 represents the plot of the data from Table 20-1. In this case, the operation could expect a $90,000 property damage accident approximately every twenty-four return periods or about once every six years. In this case the expected mean annual worst case loss would be approximately $32,000.

As with any statistical method, the information developed represents statistical probability. Even though the data developed represents a true "calculated" risk, it should be considered an order-of-magnitude estimate at best. Also, as with any risk projection technique, changes invalidate all results.

Additional information about the technique and the special paper required can be obtained from the TEAM company.

Even though extreme value projection is taught as part of the Department of Energy's MORT program, it is not required for accident investigations and is not routinely performed.

This particular technique is relatively simple and has temendous potential but, probably because of the difficulty in reproducing the special graph paper in publications, is not widely used.

TIME-LOSS ANALYSIS

Time-loss analysis is a special-purpose technique used to evaluate accident responses. System Safety Development Center courses that teach the technique credit the National Transportation Safety Board (specifically Driver and Benner) with developing and reporting the technique.

The basic premise of the technique is that for every accident there is a natural course line; that is, with no outside intervention by responding per-

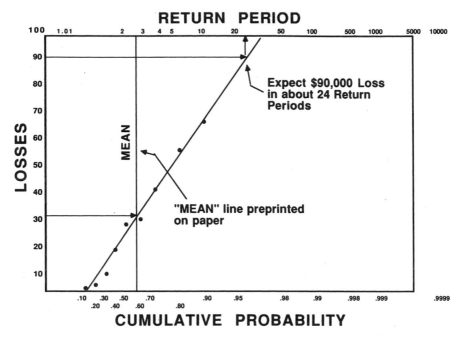

Figure 20-3 Plot of the points defined by the cumulative probabilities and losses depicted in Figure 20-3 and examples of some of the types of information that can be obtained from extreme value projection.

sonnel or systems, each accident sequence would eventually end (fires would burn themselves out, for example, or injured personnel would either recover or die). This natural course line for any given accident could be plotted by plotting time on the *x*-axis and loss on the *y*-axis (Fig. 20-4).

The effects of each intervenor could then be evaluated by examining the time at which the intervention was made during the natural course of the accident and the effect that each intervention had on losses. In all cases, intervention by emergency response personnel or others has three possible results (Fig. 20-5). The intervention typically reduces losses (for instance, the firefighters put out the fire). A second possibility, however, is that the intervention produces greater losses (a firefighter is killed in the blaze, for example, or a fire truck is destroyed by the fire). A final alternative is that the intervention has no effect on losses (the firefighters allow the fire to burn out).

This particular approach is useful in that it focuses on the effect of each intervention and not on more common evaluation criteria such as response time.

Time-loss analysis can be extremely helpful in evaluating post-accident actions. It can also be used to aid in evaluating competing courses of action, such as employing more firefighters versus installing sprinkler systems versus installing portable fire extinguishers.

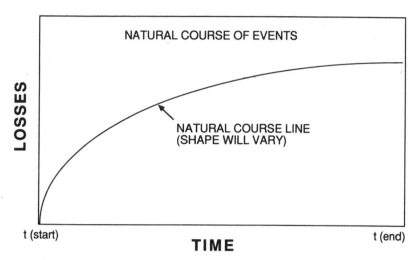

Figure 20-4 Time-loss analysis natural course line. The loss versus time plot for a given accident event with no outside intervention.

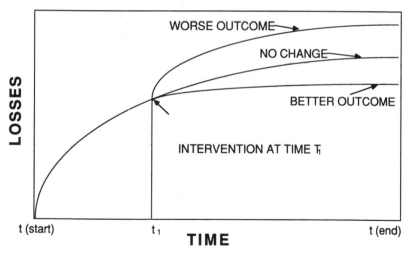

Figure 20-5 Effects of intervention. When people or systems intervene in the accident, losses may be reduced (desired outcome), losses may increase, or the intervention may not change the extent of losses.

Even though the approach is sound and can be helpful, it is sometimes difficult to apply with any accuracy of objectivity (telling what the losses are at a particular point in time while watching a fire is difficult).

Figure 20-6 is a simple example of a time-loss analysis performed for a fire in a small office building. At 8:15 A.M. a fire originates in an electrical panel in

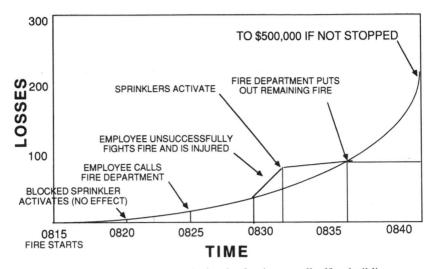

Figure 20-6 Time-loss analysis of a fire in a small office building.

a small, crowded storeroom. The only sprinkler head in the storeroom is sur-
rounded by shelves stacked to the ceiling and is ineffective but activates at
8:20. At 8:25 an employee smells smoke and calls the fire department. At 8:28
another employee locates the fire and attempts to fight it with a portable fire
extinguisher but is overcome by smoke and has to be hospitalized. By 8:32 the
fire has spread into the office outside the storeroom. At 8:34 the sprinklers in
the office activate and extinguish the fire, except in the storeroom, which continues to burn. At 8:37 local firefighters quickly extinguish the storeroom fire.
The total value of the building and contents was $500,000. The total actual
damages were $80,000, including contents of the storeroom ($10,000), water
damage ($10,000), damage to the structure ($50,000), and costs associated with
the injury to the employee who was overcome by smoke ($10,000).

ADDITIONAL TECHNIQUES

Fault Hazard Analysis

Fault hazard analysis is mentioned very frequently in system safety literature,
sometimes as a type of analysis and occasionally as a technique. One NASA
system safety document (NHB 1700.1–V3, System Safety) describes it as the
analysis to be performed after the preliminary hazard analysis for further
analysis of systems and subsystems and suggests that it can be either a separate analysis or an extension of the failure modes and effects analysis (NASA
1970). Most programs today (including NASA) refer to this analysis as the
subsystem hazard analysis (SSHA) and the *system hazard analysis* (SHA).

Gross Hazard Analysis

In some early literature gross hazard analysis described what is generally referred to as the *preliminary hazard analysis*. Like many system safety terms, it may have other meanings as well.

Hazard and Operability Studies

Sometimes referred to as *energy flow analysis*, HazOps is used primarily in the petrochemical industry. It uses a multidisciplinary team similar to a system safety working group for the systematic review of the flow of materials through a process. It concentrates on key locations in the process known as *study nodes* and uses a series of guide words and parameters to examine possible deviations and the causes and consequences of deviations (Goldwaite 1985).

Job Safety Analysis

The job safety analysis (JSA) has been a part of the industrial and occupational safety effort for many years. It is basically a method of developing job procedures that includes a systematic task analysis that examines each step of a job or task, the possible hazards associated with each step, and preventive or corrective actions required to ensure a safe operation. The technique may be appropriate to include in a life cycle system safety effort but has not traditionally been considered a system safety analysis. The JSA may be referred to as a *job hazard analysis* or *job task analysis*, even though some make distinctions between the approaches.

Technique for Human Error Rate Prediction (THERP)

One of the older and more widely used methods of addressing human performance, THERP was developed by A. D. Swain during the early to mid-1960s. It uses a six-step process to analyze and control human errors. The process involves selecting an event, identifying the tasks associated with the event, separating specific behaviors out of each task or step, assigning basic error rates based on available data, classifying the errors, and developing controls. Errors may be acts that are intentionally performed, unintentionally performed, or omitted and may occur during the input, mediation, or output phases of any given task. Like other human performance analysis efforts, the lack of accurate error rate data is a major limitation (Hammer 1972).

Cost-Benefit Analysis

Cost-benefit analysis, also called *cost-effectiveness analysis*, is used to describe the generic process of evaluating competing courses of action by examining the dollar costs of certain abatement actions versus the dollar value of the benefits received. Even though the term is widely used as a general approach, the risk management community uses cost-benefit analysis as a specific technique for evaluating insurance options (Head 1986).

Event Tree Analysis

Event tree analysis is a bottom-up, deductive technique that explores different responses to "challenges." Event tree analysis appears to be a variation of fault tree analysis. The event tree is developed from left to right, whereas the fault tree is developed from top to bottom. The initiating or "challenge" event of the event tree is similar to the undesired top event of the fault tree. Even though gates are not used in event trees, *or* gate logic is used. A primary difference is that fault trees tend to explore and list only the factors that could lead to the failure and event trees explore both success and failure alternatives at each level. Event trees can be converted to fault trees, and the procedures and formulas for quantifying each are similar (Clements 1987).

Naked Man

This technique envisions a "primitive" or unprotected system and systematically evaluates the effect of adding various controls. This approach appears to be primarily a brainstorming approach designed to detect both gaps and areas of overlapping protection to aid in identifying the most cost-effective control measures (Head 1986).

REVIEW QUESTIONS

1. What is the name of the analysis that is primarily concerned with computer codes and programs?
2. Which technique is particularly well suited for examining critical redundant components?
3. Name four parameters or characteristics that could be of concern if they are shared by critical redundant components.
4. Other than electrical, name three types of systems that may be vulnerable to sneak circuits.
5. Most sneak circuit analysis has been done by one or two large defense and aerospace contractors. Why?
6. If you are investigating an accident and the extreme value projection indicates that the accident is typical of the operation and could have been predicted from past experience, how does this result influence your investigation and your recommendations?
7. What is the recommended number of return periods to use for an extreme value projection? Why?

8. What is the primary purpose of time-loss analysis?

9. What are the three possible results of intervention by persons or systems during an accident sequence?

10. What is the meaning of a dogleg in the middle of an extreme value projection plot?

REFERENCES

Clements, P. L. 1987. *Concepts in Risk Management.* Sverdrup Technology, Inc.

Goldwaite, William H., and others. 1985. *Guidelines for Hazard Evaluation Procedures.* New York: American Institute of Chemical Engineers.

Hammer, Willie. 1972. *Handbook of System and Product Safety.* Englewood Cliffs, NJ: Prentice-Hall.

Head, George L. 1986. *Essentials of Risk ControL.* Malvern, PA.: Insurance Institute of America.

Johnson, William G. 1973. *MORT, the Management Oversight and Risk Tree.* Washington, DC: U.S. Atomic Energy Commission.

Johnson, William G. 1980. *MORT Safety Assurance Systems.* New York: Marcel Dekker.

National Aeronautics and Space Administration. 1970. *System Safety.* NHB 1700.1 (V3). Washington, DC: Safety Office, NASA.

National Aeronautics and Space Administration. 1987. *Methodology for Conduct of NSTS Hazard Analyses.* NSTS 22254. Houston: NSTS Program Office, Johnson Space Center.

Roland, H. E., and Moriarty, Brian. 1981. *System Safety Engineering and Management.* New York: John Wiley & Sons.

PART V

PROCESS SAFETY*

INTRODUCTION TO PART 5: PROCESS SAFETY MANAGEMENT

Process safety (PS) is a fascinating new and important specialization area to the safety professional and to process industry practitioners. There has always been a degree of awareness of safety related to various processes, but seldom not a focus or specific reviews for process safety throughout the life of a chemical or process industry plant.

The thrust of chemical process safety can be thought of as an extension of system safety as applied to the chemical process industry. But also important is that the safety risk-basis of the approach to achieve a less hazardous plant environment is applicable beyond chemical matters and into the facilities and operations of most industrial activities.

In this section, you will be introduced to process safety and be provided with some background, some unique definitions, an overview of the U.S. OSHA and EPA rules that are applicable to U.S. industry, and be shown the interface of system and process safety. There is also a chapter on the EPA rule that is akin to the OSHA Process Safety Management (PSM) rule. Further, there is a chapter that discusses the implementation of the PS. Finally, there is a concluding chapter that discusses PS reviews, both external and internal, to ensure that the process remains safe, but also that improvements are periodically made to enhance operations as well as safety.

This section provides a discussion about the differences between risk-based and compliance-based safety assurance and the efficiencies and advantages of having a risk-based approach.

This is a relatively new area of focus, one that is just getting of the ground, and one could easily specialize in PSM and have a career for life in the field.

This section has chapters devoted to process safety management, and in those chapters we will provide an overview of U.S. and Canadian requirements for process safety, define terms, provide some examples, and show the value of system safety philosophy when addressing the safety of processes related to the chemical industry.

---

* The information in this part is an expansion and update based on material previously provided by the author in the *System Safety Analysis Handbook* (Stephans and Talso 1997).

Chapter 21 sets the stage, defines terms, and presents the overall requirements for process safety management in the United States and Canada.

Chapter 22 describes the U.S. EPA risk management program and how it came about. It provides the public and environmental safety requirements for a facility to ensure safety beyond the fenceline of a chemical facility.

Chapter 23 discusses some suggested methods of implementation of a process safety management program in each of more than dozen requirement areas.

Chapter 24 focuses on internal and external reviews of the process safety program implementation to ensure that the level of risk to the workers and the public and environment does not increase. The reviews also verify and validate that accepted safety risk is continually minimized and that government requirements are met or exceeded. The chapter references a key inspection handbook that contains valuable information for both operations and safety personnel.

RELATION TO SECURITY

The documentation collected and developed for a PSM program is of definite interest to a person or group bent on inflicting harm to the same facility we are providing safety enhancement. Knowledge of the worst-case accidents also provides a terrorist with a "blueprint" of the vulnerabilities of a plant that we are trying to protect.

On the other hand, the information and analysis results provides the security element with a portion of an overall plant vulnerability analysis that can be used to develop a site security plan. That plan should provide actions to be taken prior to, during, and after any sort of "attack."

Security preparations should include protecting the information developed about locations, consequence, and controls; applying the safety knowledge to plant protection.

As an example, one might weigh dispersal of storage of highly hazardous chemicals rather than concentrate that storage in one potentially vulnerable location. That way, if there is a release of a smaller amount of material, the overall concentration would be less and the off-site consequence would be smaller.

There is a definite tie between process safety management and security, and in a world more sensitive to the international terrorist threat, that tie is even closer, as it should be.

CHAPTER 21

Process Safety Management

INTRODUCTION

The process safety management (PSM) of highly hazardous chemicals (HHCs) standard, 29 CFR 1910.119 (U.S. Department of Labor 1992) is intended to prevent or minimize the consequences of a catastrophic release of toxic, reactive, flammable, or explosive highly hazardous chemicals from a process. A process in this context is any activity or combination of activities including any use, storage, manufacturing, handling, or the on-site movement of highly hazardous chemicals.

The purpose of this chapter is to introduce you to process safety management, including the background that led to specific public laws, definitions applicable to this unique area, and requirements for process safety management, both in the United States and in Canada. This is a fascinating new area for both the safety or system safety professional and the process industry practitioner.

Probably the most important reason for having focus on PSM is that economies of scale dictate increasingly larger processing and manufacturing facilities and therefore the opportunity for larger disasters.

Much of the content of this chapter is based on an OSHA pamphlet (U.S. Occupational Safety and Health Department 1993) and the *System Safety Analysis Handbook* (Stephans and Talso 1997).

BACKGROUND

A host of worldwide major accidents (a chemical explosion in Texas City, Texas, in 1947 that killed 581 and injured about 3,500; the leak of methyl isocyanate gas in Bhopal, India, in 1984; and the explosion of the Piper Alpha oil platform in the North Sea in 1988; Table 21-1); focused industry and the public to the problem in the United States. This was further prompted by the vision

System Safety for the 21ˢᵗ Century: The Updated and Revised Edition of System Safety 2000, by Richard A. Stephans
ISBN 0-471-44454-5   Copyright © 2004 John Wiley & Sons, Inc.

TABLE 21-1. Significant Accidents Leading to the PSM Rule

Accident	Date(s)	Consequence	Causes	Lessons
Bhopal, India	1984	>2000 deaths	Isocyanate	Economies of scale also provide the opportunity for larger disasters. Owner Union Carbide no longer exists.
Phillips, Pasadena, TX	1989	23 deaths, 132 injuries	Explosion of flammables	A formal requirement for process safety is needed in the United States.
Piper Alpha, North Sea	1988	167 deaths	Fire/explosion of natural gas	Multiple and root cause was management related.
Seveso Italy	1974	Land contamination	Spread of dioxin	Offsite emergency coordination importance. Has now developed into a series of European Commission rules.
IMC, Sterlington, LA	1991	8 deaths, 128 injuries	Explosion of flammables	It can be expensive. IMC Global sold its entire chemicals business at a loss by the end of 2001.

that future economies of scale and encroachment on plant locations could result in larger and more catastrophic accidents.

System and Process Safety Interface

In the strictest sense, system safety encompasses process safety and the two are inextricably linked. Since systems are defined to include processes, process safety may be considered a subset of system safety. The application of appropriate system safety tenets to the chemical process industry follows. In the process industry, the system safety practitioner continues his or her quest to identify, evaluate, analyze, and eliminate or control hazards throughout the life cycle of the process (in this case). The precedence of controls remains the same. So, there is a close relation between system and chemical process safety and the process industry.

PSM Background

Risk-Based vs. Compliance-Based Requirements. Prior to the 1990s, OSHA regulations were almost exclusively compliance-based. Very specific rules were promulgated and inspections were made to ensure that the rules were followed. With the advent of the process safety regulation, a portion of it required that the risk associated with a manufacturing or other chemical site be assessed and appropriate actions be taken to mitigate the results of an accident to protect the workers.

The PSM rule provides "an integrated approach to chemical safety, putting the focus on a comprehensive management program. By integrating technologies, procedures and management practices, companies can develop a safety strategy that effectively addresses their specific processes and will help prevent potential releases of flammable gases and liquids, explosives, and pyrotechnics' (U.S. Department of Labor 1992).

Requirements

Application. The standard applies to a process that contains a threshold quantity or greater amount of a toxic or reactive highly hazardous chemicals as specified in the appendix to this chapter. Also, it applies to 10,000 pounds or greater amounts of flammable liquids and gases and to the process activity of manufacturing explosives and pyrotechnics.

Exceptions. The standard does not apply to retail facilities, normally unoccupied remote facilities, or oil or gas well drilling or servicing activities. Hydrocarbon fuels used solely for workplace consumption as a fuel are not covered, if such fuels are not part of a process containing another highly hazardous chemical covered by the standard. Atmospheric tank storage and associated transfer of flammable liquids that are kept below their normal boiling point without benefit of chilling or refrigeration are not covered by the PSM standard unless the atmospheric tank is connected to a process or is sited in close proximity to a covered process such that an incident in a covered process could involve the atmospheric tank.

Process Safety Information. Compilation of written process safety information is required, including hazard information on highly hazardous chemicals, technology information, and equipment information on covered processes.

Employee Involvement. A written plan of action must be developed regarding employee participation. Employees and their representatives must be consulter on the conduct and development of process hazard analyses and on the development of other elements of process safety management required under the rule. Employees and their representatives should have access to process hazard analyses and to all other information required to be developed under the rule. Employees include work site and contractor employees.

Process Hazard Analysis. Process hazard analyses (PrHAs) must be conducted as soon as possible for each covered process using compiled process safety information in an order based on a set of required considerations. Process hazard analyses must be updated and revalidated at least every five years and must be retained for the life of the process.

Operating Procedures. Operating procedures must be in writing and provide clear instructions for safely conducting activities involving covered process consistent with process safety information. They must include steps for each operating phase, operating limits, safety and health considerations, and safety systems and their functions. They should be readily accessible to employees who work on or maintain a covered process, and be reviewed as often as necessary to assure they reflect current operating practice. Safe work practices must be implemented to provide for special circumstances such as lockout/tagout and confined space entry.

Training. Employees operating a covered process must be trained in the overview of the process and in the operating procedures addressed previously. This training must emphasize specific safety and health hazards, emergency operations, and safe work practices. Initial training must occur before assignment. Documented refresher training is required at least every three years.

Contractors. Identifies responsibilities of work site employer and contract employers with respect to contract employees involved in maintenance, repair, turnaround, major renovation, or specialty work, on or near covered processes. Contract employers are required to train their employees to safely perform their jobs, document that employees received and understood training, assure that contract employees know about potential process hazards and the work site employer's emergency action plan, assure that employees follow safety rules of the facility, and advise the work site employer of hazards contract work itself poses or hazards identified by contract employees.

Pre-Start-up Safety Review. A safety review is mandated for new facilities and significantly modified work sites to confirm that the construction and equipment of a process are in accordance with design specifications; to assure that adequate safety, operating, maintenance, and emergency procedures are in place; and to assure process operator training has been completed. Also, for new facilities, the PHA must be performed and recommendations resolved and implemented before start up. Modified facilities must meet management of change requirement.

Mechanical Integrity. The on-site employer must establish and implement written procedures for the ongoing integrity of process equipment, particularly those components that contain and control a covered process.

Hot Work. Hot work permits must be issued for hot work operations (welding, brazing, grinding, or other fire or ignition sources) conducted on or near a covered process.

Management of Change. The work site employer must establish and implement written procedures to manage changes except "replacements in kind" to facilities that effect a covered process. The standard requires the work site employer and contract employers to inform and train their affected employees on the changes prior to start-up. Process safety information and operating procedures must be updated as necessary.

Incident Investigation. Employers are required to investigate as soon as possible (but no later than 48 hours after) incidents that did result or could reasonably have resulted in catastrophic releases of covered chemicals. The standard calls for an investigation team, including at least one person knowledgeable in the process involved (a contract employee when the incident involved contract work) and others with knowledge and experience to investigate and analyze the incident, and to develop a written report on the incident. Reports must be retained for five years.

Emergency Planning and Response. Employers must develop and implement an emergency action plan. The emergency action plan must include procedures for handling small releases.

Compliance Audits. Employers must to certify that they have evaluated compliance with process safety requirements at least every three years. Prompt response to audit findings and documentation that deficiencies are corrected is required. Employers must retain the two most recent audit reports.

Trade Secrets. Requirements exist that are similar to trade secret provisions of the 1910.1200 Hazard Communication standard requiring information required by the PSM standard to be available to employees (and employees representatives). Employers may enter into confidentiality agreements with employees to prevent disclosure of trade secrets.

International Application

While the rule applies within the United States, its provisions can be used outside its borders if the particular government or facility management is willing to adopt the rule. Adoption of the rule will certainly enhance the safety level of the process industry.

Process Safety Management in Canada

Much of the genesis of process safety in Canada originated from the United States. Proximity is perhaps the major reason, but again, the implementation of process safety makes good economic as well as safety sense. Three groups have lead the development of process safety in Canada—the Major Industrial Accidents Council of Canada (MIACC) (the original and lead organization with a charter to develop process safety), the Canadian Chemical Producers' Association (an industrial alliance), and the Canadian Society of Chemical Engineering (CSChE). Having accomplished the task provided, the MIACC was disbanded in 1999. After dissolution of MIACC, the process safety management aspect of the organization was passed to the CSChE, where is now resides.

Today, Canadian process safety management guidelines are based on those promulgated by the U.S. Center for Chemical Process Safety and are contained in the document, Process Safety Management (Center for Chemical Process Safety 1992). The dozen elements of process safety management are identified in Table 21-2 and very closely track with those of the OSHA requirement, but there are differences.

Because there are some minor differences from the OSHA rule, the components of those elements are worthy of further explanation. For example, under "human Factors" are the following components: operator–process/equipment interface, administrative control versus hardware, and human error assessment. Likewise, under "enhancement of process safety knowledge" are quality control programs and process safety, professional and trade association programs, the Center for Chemical Process Safety (U.S.) program,

TABLE 21-2. Elements of Process Safety Management in Canada

1. Accountability: objectives and goals
2. Process knowledge and documentation
3. Capital project review and design procedures
4. Process risk and management
5. Management of change
6. Process and equipment integrity
7. Human factors
8. Training and performance
9. Incident investigation
10. Company standards, codes, and regulations
11. Audits and corrective action
12. Enhancement of process safety knowledge

(Canadian Society for Chemical Engineering 2002)

improved predictive system, and a process safety resource center and reference library.

An identified stated equivalent set of guidelines for process safety is the American Petroleum Institute's Recommended Practice 750. RP750 is intended to assist in the management of process hazards. The objective of the recommended practice is to help prevent the occurrence of, or minimize the consequences of, catastrophic releases of toxic or explosive materials. The practice addresses the management of process hazards in design, construction, start-up, operation, inspection, maintenance, and modification of facilities with the potential for catastrophic release (American Petroleum Institute 1997).

Chemical Safety in Europe

In Europe, in the 1970s one major chemical accident in particular prompted the adaption of legislation aimed at the prevention and control of major accidents in the chemical process industry. The Seveso accident in northern Italy in 1976 occurred at a chemical plant manufacturing pesticides and herbicides. A dense vapor cloud containing dioxin was released due to an uncontrolled exothermic reaction. The poisonous and carcinogenic dioxin is lethal in microgram quantities and contaminated ten square miles, and more than 2,000 people were treated for exposure. Luckly, there were no immediate fatalities.

Since early 1999, the "Seveso Directive II" of the European Commission (EC) has become mandatory for industry and public authorities of the EC member states. Requirements include:

Safety management systems. The introduction of safety management systems requires development of new managerial and organizational methods in general and, in particular, significant changes in industrial practice relating to risk management.

Emergency plans. Internal emergency plans for response measures to be taken inside establishments must be drawn up by the operator and supplied to the local authorities to enable them to draw up *external emergency plans.*

Land-use planning. The provisions reflect the lessons learned from the Bhopal accident that the land-use planning implications of major accident hazards should be taken into account.

Information and public consultation. The directive gives more rights to the public in terms of access to information as well as in terms of consultation.

Accident reporting. Member states have the obligation to report major accidents to the Commission.

Inspections. In the directive, an attempt is made to ensure increased consistency in enforcement at EC level through greater prescriptive detail. An *inspection system*, which can either consist of a *systematic appraisal of each establishment* or of one *on-site inspection per year*, must be organized.

Guidance documents. In order to assist member states with the interpretation of certain provisions of the Seveso II Directive, the Commission, in co-operation with the member states, has developed the following guidance documents:

Guidance on the Preparation of a Safety Report

Guidelines on a Major Accident Prevention Policy and Safety Management System

Explanations and Guidelines on Harmonised Criteria for Dispensations

Guidance on Land-Use Planning

General Guidance for the Content of Information to the Public

Guidance on Inspections

FUTURE

In the future and within this century, several considerations related to the process industry are relatively certain, There will be continued population growth, there will be continued scientific discovery, and there will be continued innovations in technology.

With continued growth, there will not only be increased demand for product, but there will also be more of the population spreading to the locations near chemical process sites once thought to be remotely located. Product demand will also necessitate more and larger processing sites. Economies of scale (as with Union Carbide and the Bhopal plant) and the construction of even larger processing sites makes the potential for higher consequence accidents that much greater.

Scientific discovery should bring new materials, new methods, and a greater variety of facility processes. Specialization in the new areas is a likely outcome.

Innovations in technology should also foster intrinsic safety. Intrinsic safety pertains to the internal nature of a thing being safe or being free from risk that is not acceptable. From a PSM standpoint, there will be a continued quest for intrinsic safety through the use of control theory and automated process control. More information in the field is available in several texts and in trade journals such as *Control Design* (www.controldesign.com) and *Control* magazine (www.control.com).

SUMMARY

In this first chapter of Part 5, we laid the groundwork by discussing the background and accident history that led to the establishment of U.S. government process safety rules. Included was a discussion of the interface with system safety. Indeed, it can be concluded that process safety is a subset of system safety focusing on the chemical process and related industry. The chapter also highlights the requirements of the OSHA PSM rule and discusses each of its major elements. Further, there is discussion of PSM implementation in Canada and its derivation from the U.S. Center for Chemical Process Safety and other sources. Finally, the chapter addresses the future with a discussion of the trend toward further economies of scale and the expansion of populated areas to encroach on what was once considered remotely located chemical processing sites. On the positive side, there is also information related about the potential for technology-based intrinsic safety in the future.

REVIEW QUESTIONS

1. Name two "other provisions" of the PSM rule and describe the requirements of each.
2. What was the main driver for establishing the PSM rule?
3. Can other segments of industry enhance safety by using the provisions of the PSM rule? Describe.
4. Why would one consider emergency operations as a segment of system safety if it is primarily concerned with actions after a disaster?
5. While QA provisions are not specified in the PSM rule, how might they enhance the safety of process systems and operations?
6. At what stage(s) of a worker's employment is he or she required to be involved in the plant's operations and to what extent?
7. When was the PSM rule required to be fully applicable to U.S. industry?
8. State seven types of PSM PrHAs.
9. When is a worker allowed to operate a process in the chemical process industry?
10. Where are the process safety management regulations in Canada derived from?

REFERENCES

American Petroleum Institute, *RP 750 Management of Process Hazards*, First Edition, January 1990. Reaffirmed, May 1997.

Canadian Society for Chemical Engineering, *Process Safety Management*, 3rd Edition, Ottawa, Canada, 2002.

Center for Chemical Process Safety, *Plant Guidelines for Technical Management of Chemical Process Hazards*, American Institute of Chemical Engineers, 1992.

Stephans R.A. and Talso W.W., eds., *System Safety Analysis Handbook*, 2nd ed., Unionville, VA: System Safety Society, 1997.

U.S. Department of Labor News Release, 92-84, February 14, 1992.

U.S. Department of Labor, Code of Federal Regulations, 29 CFR 1910.119, *Process Safety Management of Highly Hazardous Chemicals*, 1992.

U.S. Occupational Safety and Health Department, Fact Sheet No. OSHA 93-45, *Process Safety Management of Highly Hazardous Chemicals*, January 1993.

Appendix: List of Highly Hazardous Chemicals, Toxics and Reactives

This Appendix contains a listing of toxic and reactive highly hazardous chemicals that present a potential for a catastrophic event at or above the threshold quantity.

Chemical Name	Chemical Abstract Service No.	Threshold Quantity (Pounds)
Acetaldehyde	75-07-0	2,500
Acrolein (2-Propenal)	107-02-8	150
Acrylyl Chloride	814-68-6	250
Allyl Chloride	107-05-1	1,000
Allylamine	107-11-9	1,000
Alkylaluminums	Varies	5,000
Ammonia, Anhydrous	7664-41-7	10,000
Ammonia solutions (greater than 44% ammonia by weight)	7664-41-7	15,000
Ammonium Perchlorate	7790-98-9	7,500
Ammonium Permanganate	7787-36-2	7,500
Arsine (also called Arsenic Hydride)	7784-42-1	100
Bis(Chloromethyl) Ether	542-88-1	100
Boron Trichloride	10294-34-5	2,500
Boron Trifluoride	7637-07-2	250
Bromine	7726-95-6	1,500
Bromine Chloride	13863-41-7	1,500
Bromine Pentafluoride	7789-30-2	2,500
Bromine Trifluoride	7787-71-5	1,5000
3-Bromopropyne (also called Propargyl Bromide)	106-96-7	100
Butyl Hydroperoxide (Tertiary)	75-91-2	5,000
Butyl Perbenzoate (Tertiary)	614-45-9	7,500
Carbonyl Chloride (see Phosgene)	75-44-5	100
Carbonyl Fluoride	353-50-4	2,500
Cellulose Nitrate (concentration greater than 12.6% nitrogen)	9004-70-0	2,500

Chemical Name	Chemical Abstract Service No.	Threshold Quantity (Pounds)
Chlorine	7782-50-5	1,500
Chlorine Dioxide	10049-04-4	1,000
Chlorine Pentrafluoride	13637-63-3	1,000
Chlorine Trifluoride	7790-91-2	1,000
Chlorodiethylaluminum (also called Diethylaluminum Chloride)	96-10-6	5,000
1-Chloro-2,4-Dinitrobenzene	97-00-7	5,000
Chloromethyl Methyl Ether	107-30-2	500
Chloropicrin	76-06-2	500
Chloropicrin and Methyl Bromide mixture	None	1,500
Chloropicrin and Methyl Chloride mixture	None	1,500
Commune Hydroperoxide	80-15-9	5,000
Cyanogen	460-19-5	2,500
Cyanogen Chloride	506-77-4	500
Cyanuric Fluoride	675-14-9	100
Diacetyl Peroxide (concentration greater than 70%)	110-22-5	5,000
Diazomethane	334-88-3	500
Dibenzoyl Peroxide	94-36-0	7,500
Diborane	19287-45-7	100
Dibutyl Peroxide (Tertiary)	110-05-4	5,000
Dichloro Acetylene	7572-29-4	250
Dichlorosilane	4109-96-0	2,500
Diethylzinc	557-20-0	10,000
Diisopropyl Peroxydicarbonate	105-64-6	7,500
Dilauroyl Peroxide	105-74-8	7,500
Dimethyldichlorosilane	75-78-5	1,000
Dimethylhydrazine, 1,1-	57-14-7	1,000
Dimethylamine, Anhydrous	124-40-3	2,500
2,4-Dinitroaniline	97-02-9	5,000
Ethyl Methyl Ketone Peroxide (also Methyl Ethyl Ketone Peroxide; concentration greater than 60%)	1338-23-4	5,000
Ethyl Nitrite	109-95-5	5,000
Ethylamine	75-04-7	7,500
Ethylene Fluorohydrin	371-62-0	100
Ethylene Oxide	75-21-8	5,000
Ethyleneimine	151-56-4	1,000
Fluorine	7782-41-4	1,000
Formaldehyde (Formalin)	50-00-0	1,000
Furan	110-00-9	500
Hexafluoroacetone	684-16-2	5,000
Hydrochloric Acid, Anhydrous	7647-01-0	5,000

Chemical Name	Chemical Abstract Service No.	Threshold Quantity (Pounds)
Hydrofluoric Acid, Anhydrous	7664-39-3	1,000
Hydrogen Bromide	10035-10-6	5,000
Hydrogen Chloride	7647-01-0	5,000
Hydrogen Cyanide, Anhydrous	74-90-8	1,000
Hydrogen Fluoride	7664-39-3	1,000
Hydrogen Peroxide (52% by weight or greater)	7722-84-1	7,500
Hydrogen Selenide	7783-07-5	150
Hydrogen Sulfide	7783-06-4	1,500
Hydroxylamine	7803-49-8	2,500
Iron, Pentacarbonyl	13463-40-6	250
Isopropylamine	75-31-0	5,000
Ketene	463-51-4	100
Methacrylaldehyde	78-85-3	1,000
Methacryloyl Chloride	920-46-7	150
Methacryloyloxyethyl Isocyanate	30674-80-7	100
Methyl Acrylonitrile	126-98-7	250
Methylamine, Anhydrous	74-89-5	1,000
Methyl Bromide	74-83-9	2,500
Methyl Chloride	74-87-3	15,000
Methyl Chloroformate	79-22-1	500
Methyl Ethyl Ketone Peroxide (concentration greater than 60%)	1338-23-4	5,000
Methyl Fluoroacetate	453-18-9	100
Methyl Fluorosulfate	421-20-5	100
Methyl Hydrazine	60-34-4	100
Methyl Iodide	74-88-4	7,500
Methyl Isocyanate	624-83-9	250
Methyl Mercaptan	74-93-1	5,000
Methyl Vinyl Ketone	79-84-4	100
Methyltrichlorosilane	75-79-6	500
Nickel Carbonly (Nickel Tetracarbonyl)	13463-39-3	150
Nitric Acid (94.5% by weight or greater)	7697-37-2	500
Nitric Oxide	10102-43-9	250
Nitroaniline (para Nitroaniline)	100-01-6	5,000
Nitromethane	75-52-5	2,500
Nitrogen Dioxide	10102-44-0	250
Nitrogen Oxides (NO; NO(2); N2O4; N2O3)	10102-44-0	250
Nitrogen Tetroxide (also called Nitrogen Peroxide)	10544-72-6	250
Nitrogen Trifluoride	7783-54-2	5,000
Nitrogen Trioxide	10544-73-7	250

Chemical Name	Chemical Abstract Service No.	Threshold Quantity (Pounds)
Oleum (65% to 80% by weight; also called Fuming Sulfuric Acid)	8014-94-7	1,000
Osmium Tetroxide	20816-12-0	100
Oxygen Difluoride (Fluorine Monoxide)	7783-41-7	100
Ozone	10028-15-6	100
Pentaborane	19624-22-7	100
Peracetic Acid (concentration greater 60% Acetic Acid; also called Peroxyacetic Acid)	79-21-0	1,000
Perchloric Acid (concentration greater than 60% by weight)	7601-90-3	5,000
Perchloromethyl Mercaptan	594-42-3	150
Perchloryl Fluoride	7616-94-6	5,000
Peroxyacetic Acid (concentration greater than 60% Acetic Acid; also called Peracetic Acid)	79-21-0	1,000
Phosgene (also called Carbonyl Chloride)	75-44-5	100
Phosphine (Hydrogen Phosphide)	7803-51-2	100
Phosphorus Oxychloride (also called Phosphoryl Chloride)	10025-87-3	1,000
Phosphorus Trichloride	7719-12-2	1,000
Phosphoryl Chloride (also called Phosphorus Oxychloride)	10025-87-3	1,000
Propargyl Bromide	106-96-7	100
Propyl Nitrate	627-3-4	2,500
Sarin	107-44-8	100
Selenium Hexafluoride	7783-79-1	1,000
Stibine (Antimony Hydride)	7803-52-3	500
Sulfur Dioxide (liquid)	7446-09-5	1,000
Sulfur Pentafluoride	5714-22-7	250
Sulfur Tetrafluoride	7783-60-0	250
Sulfur Trioxide (also called Sulfuric Anhydride)	7446-11-9	1,000
Sulfuric Anhydride (also called Sulfur Trioxide)	7446-11-9	1,000
Tellurium Hexafluoride	7783-80-4	250
Tetrafluoroethylene	116-14-3	5,000
Tetrafluorohydrazine	10036-47-2	5,000
Tetramethyl Lead	75-74-1	1,000
Thionyl Chloride	7719-09-7	250
Trichloro (chloromethyl) Silane	1558-25-4	100
Trichloro (dichlorophenyl) Silane	27137-85-5	2,500
Trichlorosilane	10025-78-2	5,000
Trifluorochloroethylene	79-38-9	10,000
Trimethyoxysilane	2487-90-3	1500

CHAPTER 22

EPA's Equivalent Process Safety Requirements— Risk Management Program (RMP)

While the OSHA rule applies to operations that affect workers within a site where certain highly hazardous chemicals are present at stated levels, the EPA rule was developed to additionally protect the public and the environment from the undesired consequences of explosions and other accidental releases. Much of the information in this chapter comes from DOE-STD-l100 (U.S. Department of Energy 1996) and is enhanced with associated material from the *System Safety Analysis Handbook* (Stephans and Talso 1997) and other sources as specified.

The purpose of this chapter is to describe the requirements of the EPA Risk Management Program (RMP), discuss some of the differences between it and the OSHA PSM rule, and highlight the importance of the rule to the public and the environment.

BACKGROUND

The EPA notes in its general guidance document (U.S. EPA 1998) that the RMP goal is to prevent accidental releases of substances that can cause serious harm to the public and the environment from short-term exposures and to mitigate the severity of releases that do occur.

One may ask why is this EPA rule needed if there is an OSHA PSM rule? Originally there was a parallel schedule with the OSHA PSM, but due to the tempo of enactment, the two regulations ended up being enacted about four years apart.

A major difference between the two rules is in the applicability of each. Where the OSHA rule serves to protect workers within a site, the EPA rule protects the public and the environment outside the site.

System Safety for the 21st Century: The Updated and Revised Edition of System Safety 2000, by Richard A. Stephans
ISBN 0-471-44454-5 Copyright © 2004 John Wiley & Sons, Inc.

By way of an overview introduction, the chapter will provide an overall summary of the RMP, discuss the hazard analysis requirements to include consequence analysis, describe the prevention program requirements and the emergency response requirements, and finally discuss the risk management plan.

OVERALL RISK MANAGEMENT PROGRAM

Under the rule, a risk management plan must be developed and implemented by all facilities that manufacture, process, use, store, or handle regulated substances. RMPs provide facilities with an integrated approach to identifying and managing the hazards posed by the regulated substances. A partial list of the regulated substances is provided in the appendix to this chapter. Implementation is in three levels, based on process industry type and other factors (additional implementation information is provided in Chapter 23 of this section). As shown in Figure 22-1, the RMP consists of three major parts: a hazard assessment, a prevention program, and an emergency response program. These are integrated into a risk management plan. EPA considers critical its requirement for the owner or operator of a facility to define its management system and name the person or position responsible for the program. The facility owner or operator is required to document the results of the risk management program(s) in the risk management plan. Facilities are required to maintain onsite documentation of the implementation of the risk management plan. EPA requires that the risk management plan summarize the prevention

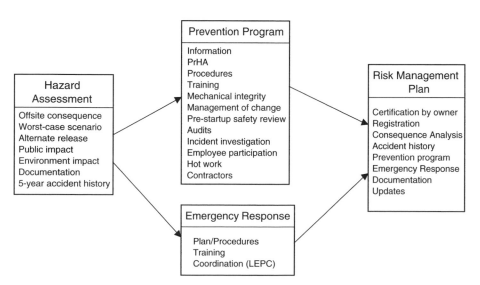

Figure 22-1 Components of a Level-3 Risk Management Program.

program elements in the plan because the information of most use to the public and local agencies is related to the hazard assessment and consequence analysis.

Hazard Assessment

The RMP rule is designed to assist facilities and communities in efforts to lessen the number and severity of serious chemical accidents. Under EPA's RMP, facilities must complete a hazard assessment to evaluate potential effects of an accidental release of any regulated substance present at or above the threshold quantity. The hazard assessment also must evaluate the impact of significant accidental releases on the public health and environment. OSHA's PSM rule requires only a qualitative evaluation of a range of possible safety and health effects on employees in the work place resulting from a release. The hazard assessment of a regulated substance requires evaluation of a range of accidental release scenarios, including:

- "Worst-case" accidental releases
- Other more probable releases
- Potential offsite consequences
- Five-year accident history for the site/acility.

The EPA rule defines "worst-case" release as the instantaneous loss of all of the regulated substance in a process, with failure of all passive and active mitigation systems. Once the worst-case and other significant accidental scenarios are identified, facilities are required to analyze the potential offsite consequences associated with these scenarios using source release and air dispersion modeling. These analyses include fires, explosions, and hazardous material releases. Air dispersion modeling is used to evaluate the fate and transport of the regulated substance for the offsite consequence analyses. At a minimum, the offsite analyses estimate the possible rate of release, the quantity released, the duration of the release, and the distances in any direction that the substance could travel before it dispersed enough to no longer pose a hazard to the public health or the environment. Along with calculating the severity of the consequences, source term modeling is used to calculate release rate as a function of time and other release characteristics.

Under the EPA rule, facilities are required to update the offsite consequence analyses of their risk management plans every five years. Updates are required sooner, if changes at the facility or its surroundings might change the results of the risk management plan to any significant degree.

Consequence Analysis

The EPA has very specific guidance to the owner or operator for the performance of offsite consequence of accidental releases of substances regulated by the Clean Air Act Amendments (and the rule). Those sites subject to the rule are required to provide results of the analyses to the state, local, and federal government and the public about potential consequences of an accidental release. The two elements of the analysis consist of a worst-case scenario and alternative release scenarios. EPA also defines the worst-case scenario as "the release of the largest quantity of a regulated substance from a single vessel or process line failure that results in the greatest distance to the endpoint. In broad terms, the distance to the endpoint is the distance a toxic vapor cloud, heat from a fire, or a blast wave from an explosion will travel before dissipating to the point that serious injuries from short-term exposures will no longer occur. In addition, EPA states that Alternative release scenarios are those "that are more likely to occur than the worst-case scenarios and that will reach an endpoint offsite, unless no scenario exists." The EPA publishes the "Risk Management Program Guidance for Offsite Consequence Analysis" (U.S. EPA 1999), which provides an EPA-approved look-up table methodology with reference tables, distance tables, and other necessary information to include an appendix of chemical accident prevention provisions. The information is free and available *via* the Internet at www.epa.gov/ceppo/.

A final element of the hazard assessment is compiling and documenting a five-year history of releases of the regulated substances. EPA's RMP requires the facility to document the releases that caused, or had the potential to cause, offsite consequences. The accident history must include:

- Substance and quantity released
- Concentration of the substance when released
- Duration of the release
- Date and time of the release
- Offsite consequence(s) (e.g., evacuations, injuries).

Note that most of the releases that meet the criteria of the EPA RMP are already reported under CERCLA and SARA Title III. Most of the information needed to define accidental release scenarios will be derived from the process hazard analysis.

As clarification of SARA Title III and CERCLA relationships, the following is provided. The Comprehensive Environmental Response, Compensation, and Liability Act (CERCLA) is designed to help clean up inactive hazardous waste sites. It also requires industries to disclose to their communities what hazardous substances they use and store. CERCLA authorized EPA to remediate polluted sites and created the Superfund to pay for site cleanups when there is no clear-cut responsible party. EPA can pursue potentially responsible parties to make them pay for response and remediation activities. Section 313 of the Emergency Planning Community Right-to-Know Act of 1986 (SARA Title III) requires EPA to establish an inventory of routine toxic chemical emissions from certain facilities. Facilities subject to this reporting requirement are required to complete a Toxic Chemical Release Inventory Form (Form R) for specified chemicals.

Prevention Program

Along with the hazard assessment and the emergency response program, the risk management plan includes a prevention program. In addition to the process hazard analysis, the prevention program covers safety precautions and maintenance, monitoring, and employee training measures. However, the OSHA PSM rule does include elements that are specific to worker protection issues that EPA has not included in its rule. Requirements of the EPA prevention program are similar to the requirements of the OSHA PSM rule with parallel elements being nearly identical. This similarity exists because EPA separates the offsite consequence analysis and the five-year accident history from the formal process hazard analysis requirements.

The integrated approach of the EPA prevention program consists of the following nine elements. These elements adopt and build on the OSHA PSM program elements:

- Process hazard analysis
- Process safety information
- Standard operating procedures
- Training
- Maintenance (mechanical integrity)
- Pre-startup review
- Management of change
- Safety audits
- Accident investigation.

EPA's RMP requires that the order in which PrHAs are conducted be prioritized based on offsite consequences. The qualitative evaluation of safety and health impacts focuses on impacts on public health and environment rather than impacts on employees. The identification of previous incidents as a part of the prevention program PrHA is limited to those with offsite consequences rather than those with catastrophic consequences in the workplace as required by the PSM Rule. Facilities are expected to have fewer incidents to consider under the EPA RMP rule, because some potential incidents will not have offsite impacts. Another EPA requirement, which is not included in the OSHA PSM rule, is that a facility define its management system. Facilities are required to identify the person (by name) or the position responsible for implementing the prevention program.

Emergency Response Program

An emergency response program also must be developed. The applicable emergency response program requirements are much more stringent under the EPA RMP than those under OSHA's PSM rule. OSHA's emergency action plan regulation basically requires an evacuation plan. Under the EPA RMP rule, facilities need to develop a more extensive emergency response plan. This emergency response plan must detail how the facility would respond to a release to limit offsite consequences. Conducting drills and exercises to test the emergency response plan(s) and coordination of plans with the Local Emergency Planning Committee (LEPC) are also required. Coordination with the LEPC is not required by OSHA although many facilities do so currently.

Risk Management Plan and Documentation

The EPA RMP rule requires facilities to prepare and submit the risk management plan to different agencies. The plan is a part of the overall RMP and the two should not be confused. The EPA also requires that a risk management plan be submitted to the Chemical Safety and Hazard Investigation Board, the State Emergency Response Commission, the LEPC, and made available to the public. The risk management plan includes the results of the prevention program elements as required, the 5-year accident history, a copy of the registration, description of the management system, and certification of the accuracy and completeness of the information. Facilities are required to maintain the documentation supporting the implementation of the risk management plan for inspection by EPA and other agencies.

Registration

The RMP rule requires facilities to register with the EPA if they have a regulated substance in a quantity greater than the threshold quantity. The rule requires facilities to register with the EPA within three years of the date of the publication of the final rule. The content of the registration includes the name and address of the facility, the facility's Dun and Bradstreet number, the regulated substances on site and the quantities, and the facility's Standard Industrial Code (SIC) that applies to the use of each regulated substance. Most of these registration requirements are already reported under the SARA Title III. If the information on registration changes (e.g., quantity change, new chemicals added, chemicals are no longer used) after the submittal of the registration, facilities are required to file an amended registration form within 60 days.

SUMMARY

The EPA's Risk Management Programs for Chemical Accidental Release Prevention (40 CFR Part 68) add significant requirements beyond those in

OSHA's "Process Safety Management of Highly Hazardous Chemicals" (29 CFR 1910.119). In addition, the chemical list and threshold quantities for the EPA rule differ somewhat from those in the OSHA rule and result in a facility needing to expand their risk management program to other portions of the facility. The three principal areas in which the requirements of the EPA exceed those of the OSHA Rule are:

1. Performance of hazard assessments that includes analyses of the "worst case" accident consequences.
2. Preparation of a written risk management plan to document the risk management program. The plans are submitted to designated agencies and available to the public.
3. Registration of the risk management plans with the EPA.

REVIEW QUESTIONS

1. What is the physical location where EPA jurisdiction begins related to the RMP?
2. List some of the differences between the PSM (29 CFR 1910.119) rule and the PSM rule (40 CFR 68).
3. Why was the RMP needed when there was already a PSM rule?
4. Define the endpoint quantity as it relates to the RMP consequence analysis.
5. What is one of the best sources of information about implementation of an RMP?
6. What must be included in the five-year site accident history?
7. What information is required in the EPA's RMP registration form?
8. Discuss the security implications of the RMP information made available to the public.
9. Why would coordination with the Local Emergency Planning Committee be required under the RMP and not be required under the OSHA PSM rule?
10. How many different flammable substances above the threshold quantity might trigger compliance with the RMP?

REFERENCES

Stephans R.A. and Talso W.W., eds., *System Safety Analysis Handbook*, 2nd ed. Unionville, VA: System Safety Society, 1997.

U.S. Department of Energy, DOE-STD-1101-96, *Process Safety Management for Highly Hazardous Chemicals*, February 1996.

U.S. Environmental Protection Agency, EPA 550-B-98-003, *General Guidance for Risk Management Programs (40 CFR 68)*, July 1998.

U.S. Environmental Protection Agency, EPA 550-B-99-009, *Guidance for Offsite Consequence Analysis*, April 1999.

Appendix: Seventy-six Substances Listed Under 40 CFR 68

Process/facility quantities greater than the threshold quantity invoke Risk Management Program requirements. (Note that there are an additional 77 flammable substances not included here.)

Chemical Name	Threshold Quantity (TQ in pounds)
Acrolein	5,000
Acrylonitrile	20,000
Acrylyl chloride	5,000
Allyl alcohol	15,000
Allylamine	10,000
Ammonia (anhydrous)	10,000
Ammonia (>20%)	20,000
Arsenous trichloride	15,000
Arsine	1,000
Boron trichloride	5,000
Boron trifluoride	5,000
Boron trifluoride with methyl ether	15,000
Bromine	10,000
Carbon disulfide	20,000
Chlorine	2,500
Chlorine dioxide	1,000
Chloroform	20,000
Chloromethyl ether	1,000
Chloromethyl methyl ether	5,000
Crotonaldehyde	20,000
Crotonaldehyde (E)	20,000
Cyanogen chloride	10,000
Cyclohexylamine	15,000
Diborane	2,500
Dimethyldichlororosilane	5,000
Dimethylhydrazine	15,000

Chemical Name	Threshold Quantity (TQ in pounds)
Epichlorohydrin	20,000
Ethylenediamine	20,000
Ethyleneimine	10,000
Ethylene oxide	10,000
Fluorine	1,000
Formaldehyde (solution)	15,000
Furan	5,000
Hydrazine	15,000
Hydrochloric acid (>37%)	15,000
Hydrocyanic acid	2,500
Hydrogen chloride (anhydrous)	5,000
Hyrdogen fluoride/Hydrofluoric acid (>50%)	1,000
Hydrogen selenide	500
Hydrogen sulfide	10,000
Iron, pentacarbonyl	2,500
Isobutyronitrile	20,000
Isopropyl chloroformate	15,000
Methacrylonitrile	10,000
Methyl chloride	10,000
Methyl chloroformate	5,000
Methyl hydrazine	15,000
Methyl isocyanate	10,000
Mrthyl mercaptan	10,000
Methyl thiocyanate	20,000
Methyltrichlorosilane	5,000
Nickel carbonyl	1,000
Nitric acid (>80%)	15,000
Nitric oxide	10,000
Oleum	10,000
Peracetic acid	10,000
Perchloromethylmercaptan	10,000
Phosgene	500
Phosphine	5,000
Phosphorous oxychloride	5,000
Phosphorous trichloride	15,000
Piperidine	15,000
Propionitrile	10,000
Propyl chloroformate	15,000
Propyleneimine	10,000
Propylene oxide	10,000

Chemical Name	Threshold Quantity (TQ in pounds)
Sulfur dioxide (anhydrous)	5,000
Sulfur tetrafluoride	2,500
Sulfur trioxide	10,000
Tetramethyllead	10,000
Tetranitromethane	10,000
Titanuim tetrachloride	2,500
Toluene 2,4-diiosocyanate	10,000
Toluene 2,6-diiosocyanate	10,000
Toluene diiosocyanate (unspecified isomers)	10,000
Trimethylchlorosilane	10,000

■■■■■■■ CHAPTER 23

Process Safety Implementation

INTRODUCTION

In this chapter, we will additionally discuss PSM rule provisions and some thoughts about the method of implementation of it and the RMP rule. Again, much of the information in this chapter comes from the DOE-STD-1100 (U.S. Department of Energy 1996) and it is enhanced with associated material from the *System Safety Analysis Handbook* (Stephans and Talso 1997) and other sources as referenced.

This chapter addresses each of the 14 elements of the OSHA PSM rule and provides implementation thoughts and suggestions for each. The chapter additionally has a section about EPA RMP implementation that varies based on the results of the consequence analysis or dependent on the type of process industry.

PSM IMPLEMENTATION

The very first step to take in the implementation of any program is to research the issue. Does the PSM rule even apply to the site and, if so, to what facilities/processes? The PSM rule is extensive and implementing it requires scoping, planning, and scheduling of actions. Then, the next step and prior to getting into the PSM implementation details involves preparing that plan of action, identifying resources needed for accomplishment. Finally, and prior to starting, the plan needs to be approved and coordinated. Only then, should one begin the implementation detail. That detail is shown in Figure 23-1 and discussed in the next several pages. When fully implemented throughout the country, the PSM rule is estimated to prevent more than 260 deaths and 1,500 injuries annually (U.S. Department of Labor 1992).

System Safety for the 21ˢᵗ Century: The Updated and Revised Edition of System Safety 2000,
by Richard A. Stephans
ISBN 0-471-44454-5   Copyright © 2004 John Wiley & Sons, Inc.

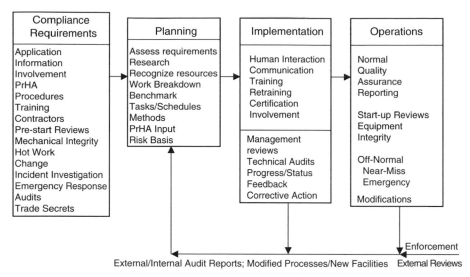

Figure 23-1   Closed-Loop Process Safety System.

1. Employee Participation

The contractor/owner should develop a plan that provides for an appropriate level of employee participation in the conduct of each required PSM element:

- Consultation with workers on the PrHA
  - Employee access to PrHAs and process safety information
  - Training
  - Participation in the investigation of mishaps and issuance of the results of the investigation

2. Process Safety Information (PSI)

The contractor/owner must assure that there is adequate information available to support the preparation of PrHAs, operating procedures, training materials, and emergency plans. The PrHA team can generate information on process technology when complete information does not exist. In addition, documentation must be available to affirm that the process equipment complies with recognized and generally accepted good engineering practices. Some of the more frequently used codes and standards include:

American National Standards Institute (ANSI)
American Society for Testing and Materials (ASTM)
National Fire Protection Association (NFPA)
American Society of Mechanical Engineers (ASME)

Institute for Electrical and Electronics Engineers (IEEE)

American Petroleum Institute (API)

For existing equipment designed and constructed according to codes, standards, and good practices no longer in use, a determination must be made and documented that the equipment is designed, maintained, inspected, tested, and operated in a safe manner. Consider the following methods when determining that the existing equipment and systems demonstrate that they are designed for safe operation:

1. Conduct analyses or empirical testing to demonstrate that the equipment/systems design provides a level of protection equivalent to current codes and standards.
2. Change process parameters to comply with current codes or standards is more conservative.
3. Use the PrHA to demonstrate that continued use of existing equipment does not significantly increase the likelihood of catastrophic consequences, compared with equipment designed to current standards.

Information about the process must include the information required by the OSHA rule to include data on the chemistry and chemical hazards, the process methodology, and the process equipment. This includes chemistry and chemical hazards information to include physical data, toxicity, permissible exposure limits (PELs), reactivity data, thermal and chemical stability, hazardous effects of mixing with other chemicals, and corrosivity data.

Information on the process technology must include process flow diagrams; process chemical engineering; maximum intended inventory in various locations; safe limits of pressure, temperature, flow, levels, phases, and composition for all modes; and consequence of deviations outside the safe limits.

Information about process equipment including materials of construction, electrical classification, relief systems, HVAC, design codes and standards, material and energy balances, safety systems, and piping and instrumentation diagrams (P&IDs) must also be identified and assembled.

3. Process Hazard Analysis (PrHA)

A team of two to five members who have expertise with engineering and process operations, specific process experience, and a leader with knowledge of methodology is required to perform the PrHA.

The team must use appropriate hazard analysis methodology and consult with those involved with the process operation and maintenance. Specified methods include "what-if", checklist, "what-if"/checklist combination, failure modes and effect analysis, hazard and operability study, fault tree analysis, or an equivalent method. The team must identify process hazards; review the accident history to identify hazards, accident precursors, lessons learned, and trends; consider the impacts of human factors; identify engineering and administrative controls and their interrelationships; determine worst-case consequences; and determine the range of safety and health effects on workers.

On completion of the PrHA, appropriate findings and recommendations for corrective action are usually presented. These recommendations must be reviewed and acted upon by plant management. It is recommended that, if there are many recommendations, they be entered and tracked with a database tracking system. Management may only reject some of the findings on the basis of factual errors, not necessary for the protection of health and safety, an alternative measure providing a sufficient level of protection, and the recommendation being infeasible.

4. Operating Procedures

Operating procedures must be prepared to cover all phases of the process to include start-up, normal or partial operation, temporary operation, normal shutdown, and start-up following a turnaround or after and emergency shutdown. They must also address off-normal actions. That is, procedures must include operating limits and the steps taken to correct or avoid deviation from those limits and operators must know the consequences of deviations, what actions to take, and how to use safety equipment.

When changes to procedures are requested, they must have a safety review prior to incorporation. The procedure must also be reviewed each time there is a change (even a subtle change) to the process equipment.

Industrial safety, quality, emergency management, and security procedures interface the operating procedures and they need to be reviewed and referred in the development of the operating procedures.

Operating procedures are the basis for operator training as discussed in the next section. They need to be where the operators can refer to them, and an additional control copy should be available in the control room of the plant.

5. Training

Initial and refresher training every three years or more frequently is required and there should be a training procedure to state the PSM training method.

The initial training includes safe work practices, the hazardous materials of the process and safety measures, emergency procedures, operating procedures and job hazard analyses, accidents and near misses, operating limits and actions and the results of deviation, and equipment and process parameters.

Refresher training needs to focus on current operating procedures, changes in equipment or processes, impacts of recent process or equipment changes, and process accidents or near misses.

Written examinations, although not required, are suggested as a means to assure that training is effective.

6. Subcontractor Safety

While it is important to ensure that subcontractors work safely, it is most important for the owner or prime contractor to understand the overall responsibilities vis-à-vis subcontractors. Those responsibilities include:

- Obtaining safety-related information about the subcontractor such as the latest OSHA 200 Log, current health and safety information, and safety and related procedures and manuals. This information is best requested in the statement of work for job that is being contracted.
- Informing subcontractors about the processes and of the hazards related to the work to be done.
- Providing the subcontractors a copy of the PrHAs and process safety information related to the work to be accomplished.
- Explaining the applicable provisions of the emergency plan to the subcontractors.
- Developing and implementing safe work practices to control the entrance, exit, presence, and accountability of subcontractors.
- Periodically evaluating subcontractor safety performance and compliance with the PSM requirements.
- Verifying that each of the subcontractors on the job is properly trained to perform the work.
- Maintaining an illness/injury log.
- Establishing a work authorization or permit system to ensure that the presence of subcontractors in the vicinity of PSM processes is known.
- Providing information from the relevant portions of process information and emergency requirements to subcontractors for their protection.

Subcontractors, on the other hand, in order to perform their tasks in a safe and coordinated manner must:

- Train their personnel to the fire, explosion, or toxic release hazards related to the job and the provisions of emergency plans.
- Document worker successful completion of training to include employee names and date of such training.
- Ensure that the subcontractor's workers follow the safety rules of the plant.
- Inform the owner or prime contractor of any unique hazards of their work or any hazards found that were not previously reported.

Only when there is an adequate system for the above responsibilities can the requirements be considered as implemented.

7. Pre-Start-up Safety Review

Much like a PrHA, the pre-start-up safety review is conducted by a team of experienced and knowledgeable personnel that examine the physical facility, associated documentation to include procedures, and proposed workers. No live processing is accomplished prior to the Pre-start-up safety review.

A tracking system should also be in place to follow the findings and recommendations until a resolution has been achieved. The process should not be operated with unresolved pre-start-up safety review findings, issues, or recommendations.

The pre-start-up safety review should be in the level of detail to assure safe operation and there should be a pre-start-up safety review plan of action should be developed and approved by management.

8. Mechanical Integrity

The proper maintenance and quality assurance tasks performed for design, construction, procurement, and repair of process equipment is vital to the safe operation of the facility. There should be a program to ensure the quality of the material coming into the facility—material for construction as well as process materials and chemicals. Likewise, during daily operations, if there are deviations from safe operation that require corrective maintenance, the operation must be stopped until such repairs are made.

Additionally, the following should be a part of the mechanical integrity system of the facility:

- Written maintenance procedures.
- Inspection and test procedures.
- Trained process maintenance personnel.
- Scheduled inspection, testing, and maintenance/replacement of process and supporting equipment and materials.
- A QA program to ensure that new construction, materials, or equipment is according to design requirements and that installation and construction is properly accomplished. QA also ensures that proper replacement parts are received and installed in accordance with design/manufacturer's requirements.
- There is an adequate preventive/predictive maintenance program.

The preventive maintenance/predictive program covers pressure vessels and storage tanks, valve and piping systems, relief and vent systems, emergency shutdown systems, fire suppression systems, controls, and pumps.

Qualified and trained workers must be used for maintenance and quality assurance tasks and if subcontractors' work must be assured of being the same quality level.

9. Nonroutine Work Authorizations (Hot Work)

Again, hot work is that type of work that involves welding, grinding, cutting, and other flame- or potential spark-producing activity that might be performed in the vicinity of processes covered by the process safety management rule. The hot work permits provide for fire safety in conjunction with that type of work and typically include written authorizations and procedures for special controls and surveillance requirements. Most important is the coordination of maintenance and operations.

For the hot work permitting process, management must establish areas for hot work, designate the authorizing individual, ensure that the personnel performing the work are trained, and ensure subcontractors who perform the work have similar requirements. From personal experience, it is especially wise with welding operations to have a fire watch. This is a person who can watch for sparks and small fires that may not he observed by the welder, who is required to wear eye protection.

The hot work permits must have the date for the work, identify the equipment on which the hot work is to be performed, identify locations where sparks may drop, describe the location and type of fire extinguishers, assign fire watchers, state fire precautions, identify prohibited areas, require relocation of combustibles, and identify the need to shut down any systems that may be a potential fire hazard.

10. Management of Change

A simple modification in a valve on a process might well have significant safety implications and require changes in the normal and emergency procedures, maintenance inspections and schedules, and worker training. Therefore, the implications of process changes must be recognized and understood.

Just as a formal program is needed for engineering design changes prior to construction, a similar program is needed to identify and track changes for operations during and after construction.

The written management of a change program or procedure must address responsibilities for approval, basis for change, the method to initiate a change, change tracking, and criteria for change approval. There should be documentation for the change basis and an acceptable analysis of risk and impact associated with the proposed change prior to the change being approved. OSHA provides a "Request for Change" form in its Pamphlet 3133 (U.S. Department of Labor, OSHA 1992).

If and when changes are approved, there should be a positive mechanism that will ensure that there are the required resources to effect the change, the process safety information and PrHA are appropriately updated, there ate updated procedures (operating and emergency) and training, that the personnel who may be affected are informed, that there is adequate pre-start-up review, and that there is a required level of maintenance and quality assurance.

In addition to the change control tracking system, there should also be some formal system to periodically review and audit the management of change program for effectiveness and continuous improvement.

11. Incident Investigation

Investigations of mishaps that could or do lead to a catastrophic release of hazardous chemicals are mandatory. Most important is to investigate occurrences that *could* lead to release. These are more subtle, but important nonetheless, and may signal an eventual catastrophic release if no further action is taken.

Since the investigation must be initiated within 48 hours, it is important to have investigators or an investigation team trained and designated prior to any occurrence. That team should have an experienced leader, additional technical expertise, and a person knowledgeable of the process involved in the mishap. Data and records analysis, information review, reconstruction, interviews, equipment examination, and continuous team consultation are important to efficient closure. An "Accident Investigation Report" form is included in the OSHA Pamphlet 3133 (U.S. Department of Labor, OSHA 1992).

Root cause analysis is a most important technique that can be used in determining the direct cause and contributing factors for the mishap. Succinctly, the technique asks successive questions of "why did this event happen?" until no further response is available and the root cause should be identified.

Implementation of investigation findings and recommendations must be approved by management and performed in a timely manner. Feedback from the investigation should be integrated into process improvements, training, and procedures and the report should be available to those who are affected, especially workers.

12. Emergency Planning and Response

While most chemical facilities have an emergency plan, it may not be as detailed as required for the PSM rule and it may not have addressal of the response and recovery stages of an emergency. The plan is required to include: types of accidents considered, emergency procedures and responsibilities, escape routes, locations for safe zones (assembly points), types of alarms, actions before evacuation, personnel accounting, rescue and medical responsibilities, reporting, local coordination, and the procedures for handling small releases of hazardous material.

Adequate and quick response to an emergency can preclude more serious consequences. Key actions after the emergency is declared to limit further undesired consequence may include safe operations shutdown, rescue, firefighting, activation of emergency communication systems, and external plant support notification.

13. Compliance Audits

Each portion of the PSM rule implementation must be audited every three years by the employer (owner/operator) of the facility and there should be a procedure in place that describes the overall audit process.

Internal Audits. Formal internal reviews or audits are required by the PSM rule every three years by a person knowledgeable of the process. The purpose of the audit is to determine compliance with the PSM rule and therefore, each of the 14 requirements are examined for proper implementation. It is presented to the owner as a written report.

The checklist in the OSHA Inspection Manual (see Chapter 24) is probably the best comprehensive tool that can be used to evaluate the process for compliance. Use of the checklist is probably the best way of assuring that initial and subsequent implementation of PSM has been accomplished. The process of assessing each provision of the rule with the checklist will yield findings regarding the adequacy of implementation. Each of the formal audit findings must be promptly responded to by the employer and correction of deficiencies must be documented.

Implementation. The owner/operator must certify on the basis of the compliance audit every three years that the process meets the requirements of the PSM rule and retain that certification for OSHA inspection as an official record. External audits by OSHA in the United States are occasionally performed and are prompted principally by a reported problem or problems at a plant location. More information about the internal and external audits is provided in Chapter 24 that follows later in this part of the book.

14. Trade Secrets

Proprietary information and trade secrets are often associated with the technical design and construction of chemical processes. Those associated with the processes and who have a need to know the sensitive information must be trusted not to divulge that information outside the organization without authorization. In most cases, to provide protection, each employee (or subcontractor) is required to sign legally binding statements that he or she will keep the information in confidence and protect it.

RMP IMPLEMENTATION

While the ultimate objective is increased safety and less chance for process accidents, the application and implementation of the RMP differs from that of the PSM rule. General guidance on RMP implementation is available in the EPA's RMP series as document EPA 550-B-98-003 (U.S. EPA 1998). Like the PSM rule, there is a list of hazardous chemicals with threshold quantities that trigger implementation. However, there are also three levels of implementation for processors of the hazardous chemicals depending on the type of industrial process and prior accident history. Figure 23-2 highlights the triggers for the each of the three levels. Key to the designation is the result of the consequence analysis.

Consequence Analysis

According to the EPA's guidance (U.S. EPA 1998), those sites subject to the rule are required to conduct an offsite consequence analysis and provide the information to state, local, and federal governments and to the public about potential consequences of an accidental chemical release. The offsite consequence analysis has two potential elements: 1) worst-case release scenario and 2) alternate release scenario. The EPA defines worst-case scenario as the release of the largest quantity of a regulated substance from a single vessel or process line failure that results in the greatest distance to an endpoint. As noted in Chapter 22, distance to endpoint is the furthest distance the effects from short-term exposures will no longer occur. A table of toxic endpoints is provided in Appendix A of 40 CFR 68 and the quantity is provided in milligrams per liter. Alternative release scenarios that are more likely to occur

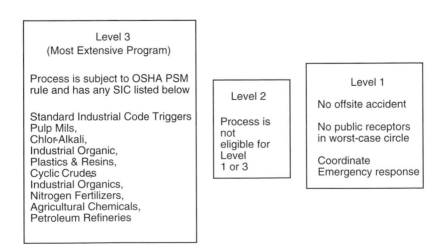

Figure 23-2   RMP Program Implementation Tiers.

than the worst-case scenario and that will reach an endpoint offsite, unless no such scenario exists.

To assist with the planning for consequence analysis, the National Oceanic and Atmospheric Administration have developed software that performs the calculations. The software can be downloaded from the Internet and its application is described in the EPA reference (U.S. EPA 1998). The table of listed substances and toxic endpoints is included in the software program.

Within the two types of scenario parameters, the site has flexibility to choose the type of release scenarios appropriate for the consequence analysis.

Implementation Degree

With the results of the consequence analysis, the site can then verify the level or degree of implementation required by the site owner or operator. It is important to also realize that over time and with changes to the site, the consequence analysis may require or allow for level of implementation adjustments (Table 23-1).

Since the provisions of the PSM rule very closely follow that of the RMP, it is emphasized that the two programs be integrated for more efficient implementation and oversight.

TABLE 23-1. RMP Requirements for Implementation Levels

Requirements	Level 1	Level 2	Level 3	Comments
Consequence Analysis	X	X	X	Applies to all. Alternate release scenario does not apply to Level 1.
Management Program		X	X	Does not apply to Level 1. Similar to that of the PSM rule.
Emergency Response	X	X	X	Level 1 requires local responder coordination. Level 2 & 3 additionally require a plan and program.
Prevention Program		X	X	More detailed program at Level 3. Similar to the multiple areas of the PSM rule.
Risk Management Plan	X	X	X	Content same for Levels 2 & 3. Less for Level 1. All require 5-year accident history.
Registration	X	X	X	Applies to all. Relatively simple form sent to local, state, federal governments and available to public.

IMPLEMENTATION LESSONS

In part, continuous improvement can be achieved by learning and recording the lessons from those who have done similar work in the past. Note that many are repeats of previous lessons. They are listed because of their significance and the need not to repeat any correctible shortcomings of the past. Such lessons from process hazard analysis activities are:

- Have a plan and use it. Planning ahead is always required, but too often the plan gathers dust until just before a major due date. Setting of intermediate goals helps to divide up the tasks for major deliverables.
- Develop the path forward tools. This may include a format and content guide and a project style guide early. The tools help to have a unified direction for each of several analysis individuals or teams to follow and it reduces the criticism of having an interim product that is inconsistent.
- Compile reference material early. The sooner all of the pertinent data and documents are assembled, the sooner a meaningful evaluation can be conducted. If there are delays in securing key materials, there is the danger of having a false start.
- Coordinate on the method. Ensure that the customer and the customer's customer agree on the evaluation methods to be used as well as the method of presenting results. There should be a clear understanding and acceptance of the specific hazard analysis process with management and the regulator. Again, this helps to assure that the effort is coordinated and approved ahead of time.
- Provide continuous update. Ensure that those in the management chain are provided frequent status reports to include any concerns for the future.
- Ensure attention to detail. Usually, if the details are all addressed, the big issues take care of themselves. Especially when in some cases there are more reviewers than analysts performing the work, this is a well-learned lesson.
- Focus on factual, not subjective. In some instances, factual results can seem counterintuitive, but it is factual material that supports good evaluation.

PSM Subterfuge

The PSM rule tends to ensure that sites are operated in a safe manner by imposing a set of rules that may be interpreted as burdensome. Even properly implemented, the rules do not guarantee an accident-free site. Nevertheless, in order not have to comply with the "burdensome" rule, some owners divide the quantity of chemicals that would normally trigger compliance with the rule into amounts under the threshold quantity. As an example, the threshold amount of chlorine at a site location that would trigger compliance with the PSM standard is 1500 pounds. Rather than using a 1-ton container, a site could use multiple smaller gas cylinders and store them at several different locations on the same site. When a cylinder needs replacement, another is moved to the use point. While this is quite legal, the site is not benefiting from the more adequate level of overall safety in implementing the PSM rule.

SUMMARY

Proper implementation of process safety requirements is key to safety assurance at covered process sites. In this chapter we attempted to identify actions prior to and during the implementation of the PSM and RMP process safety programs. Specific actions for each of the 14 PSM elements were highlighted as well as a section on RMP implementation. Finally, a section on lessons learned and information regarding implementation evasion was presented.

REVIEW QUESTIONS

1. Prior to undertaking the details of process safety implementation what actions are needed?
2. Name five recognized organizations that promulgate standards for design.
3. Why is the PrHA such a fundamental and important document?
4. Discuss why, if engineered controls are so important when compared with administrative controls, there seems to be emphasis on these in both the PSM and RMP programs.
5. Should change management be a formalized process and, if so, why?
6. Discuss how the owner/operator of a site knows that the site processes are ready for operations in the context of PSM/RMP.
7. What are the main bases for RMP implementation?
8. Generally, what are the differences between the three implementation levels for RMP?
9. Name three lessons learned associated with process safety.
10. Is the solution to pollution necessarily by dilution? Explain with reference to evading PSM and/or RMP requirements.

REFERENCES

Stephans R.A. and Talso W.W., eds., *System Safety Analysis Handbook*, 2nd ed. Unionville, VA: System Safety Society, 1997.

U.S. Department of Energy, DOE-STD-1101-96, *Process Safety Management for Highly Hazardous Chemicals*, February 1996.

U.S. Department of Labor News Release, 92-84, February 14, 1992.

U.S. Department of Labor, Occupational Safety and Health Administration, Pamphlet 3133, *Process Safety Management Guidelines for Compliance*, 1992.

U.S. Environmental Protection Agency, EPA 550-B-98-003, *General Guidance for Risk Management Programs (40 CFR 68)*, July 1998.

CHAPTER 24

Process Safety Reviews

INTRODUCTION

In everyday life, while we usually trust people who work for us to perform professionally and do their jobs, we also check that those job tasks are done correctly and to our satisfaction. That's what audits are about, only in a more formal manner. This chapter discusses the Process Safety Management (PSM)/Risk Management Program (RMP) requirement (U.S. Department of Labor 1992; U.S. EPA 1996) to perform audits of covered processes to ensure that they are adequately operating. After the basics are defined, after requirements are understood, and after the PSM/RMP system has been developed, it is time to perform internal reviews and prepare for external reviews.

Topics addressed in this chapter include audit purpose, QA interface, internal and external audits, conduct of audits, lessons learned, and information about the Chemical Safety and Hazard Investigation Board. Conduct of audits will include a section on the mechanics of an individual audit that should be most helpful during the periodic review process.

Purpose

Audits or some equivalent means of operational reviews are an important segment of any quality assurance program. They provide feedback to management and are the means of ensuring a continuous improvement of the system and overall facility. In the context of audits for process safety there is a close linkage with quality assurance. Quality assurance encompasses the spectrum of not only ensuring that the requirements are met, and it also provides a system to achieve continuous improvement, in addition to compliance.

System Safety for the 21st Century: The Updated and Revised Edition of System Safety 2000, by Richard A. Stephans
ISBN 0-471-44454-5   Copyright © 2004 John Wiley & Sons, Inc.

QA Component of Safety

Quality and safety are mutually supportive in providing overall site and/or system enhancement. QA enhances safety in several ways. First, the pedigree of materials associated with and that go into the process and facility should have acceptable attributes to ensure fitness for use and reliability under the full operating environment and any upset environment. Secondly, QA audits provide feedback that point out potential problems or concerns before they rise to a point causing a mishap. In addition to identifying issues to be corrected, the audits surface ideas for improvement of the process and the site. QA supports safety in a number of other ways to include the contracting function, design, and management aspects. Similarly, enhanced safety and feedback mechanisms tend to provide for enhanced product quality, less downtime, and more efficient conduct of operations. As an example, PSM audit aspects focused on maintenance operations and process/equipment integrity necessarily examine the preventive, predictive, and emergency maintenance process within the site to ensure the system is adequate. By doing so there is enhanced quality.

Benchmarking

A QA technique that is helpful with process audits is a means to determine the standard of compliance with the regulation. Benchmarking defines the industry standards of excellence in a particular area. In the case of the process industry, there are several sources to determine the benchmark. Such sources as technical journal articles, industry associations, personal contacts, and the industry itself are often as close as a search engine for the internet. Google or Yahoo might be initial starting points.

Internal Audits and Reviews

In order to assure the provisions of the PSM rule are properly implemented, some degree of independent review is appropriate. Additionally, reviews should be periodically accomplished to ensure that with time, standards of implementation are maintained. Whether they are called reviews, audits, surveys, or inspection is not as important as the fact that they are being conducted. Too many times, a site feels that operations are being conducted "by the book" when at closer observation, they are not. The purpose then is to double-check that the status is as advertised.

External Audits and Reviews

While external audits, reviews, or assessments are usually thought of as state or federal government reviews, they may not be. Especially in larger corporations, there is usually some independent audit function that performs external reviews. The general QA reviews measure item and service quality, adequacy of performance, and they promote improvement. Key to review performance is to ensure the persons who perform the reviews are technically knowledgeable and qualified in the areas to be reviewed.

Conduct of Audit Operations

According to the PSM rule, employers must certify that they have evaluated compliance at least every three years to verify that the procedures and practices developed under the standard are adequate and are being followed. Audits are to be performed by at least one person who is technically familiar with the process.

Especially in the nuclear industry and other potentially high consequence endeavors, there are formal sets of rules for the QA and safety audit process. ANSI/ASME NQA-1 (ASME 2001) nuclear quality audits use certified and trained auditors and lead auditors who are also expert in the areas being audited.

Nuclear Process Industry

Because of the potential for high-consequence accidents, since inception the nuclear industry has been closely regulated and audited by the Nuclear Regulatory Commission. The audit process is formalized with the industry standard ASME/ANST NQA-1 (ASME 2001). The NQA-1 standard provides comprehensive QA requirements and audit process requirements to include auditor and lead auditor training, certification, and continuance of certification standards.

OSHA Model

The OSHA has formalized the process of audits in a comprehensive manner. The process is documented in OSHA Instruction CPL 2-2.45A (U.S. Department of Labor 1994), which hereafter will be referred to as the OSHA Inspection Manual. The Manual, which is available on the web is most important for the PSM safety practitioner to have and is considered an essential reference in preparing for external compliance reviews of PSM implementation. It is also most helpful during the PSM implementation process. For each of the 14 PSM requirements, the OSHA Inspection Manual provides inspection information to include an overview of the program elements, references, verification of records direction, on-site review method, and interview coverage, so that there is consistent and adequate review of each inspected site. The Manual establishes uniform OSHA inspection policies, procedures, standard clarifications of PSM requirements, and provides compliance guidance for enforcement of the PSM rule. It also provides amendments to the rule for the Explosives and Blasting regulation, 29 CFR 1910.109. It is interesting to note that an earlier

TABLE 24-1. OSHA Inspection Manual Contents

Manual Contents
Background
Types of OSHA Inspections
Inspection Resources
Program Quality Verification (PQV) Inspection Scheduling
Scope of PQV Inspections
PQV Inspection Procedures
Compliance Guidelines for Specific Provisions of 29 CFR 1910.119 Citations
APPENDICES:
PSM Audit Guidelines and 59 pages of Checklists
Clarifications and Interpretations of the Standard
Standard Industrial Classification (SIC) Codes Targeted for PQV Inspections
References for Compliance with the Standard
(Reserved)
Sample Letters
Recommended Guidelines for PQV Inspection Preparations

version (1988) of the document was titled, "Systems Safety Evaluation of Operations with Catastrophic Potential." Table 24-1 below provides an outline of the OSHA Inspections Manual.

The Manual contains 59 pages of checklist questions to solicit a "yes" or "no" response from site documentation reviewed, activities observed, and/or personnel interviewed. According to the Manual, "Any 'no' response shall normally result in a citation for violation of that provision" (U.S. Department of Labor 1994).

MECHANICS OF AN INDIVIDUAL AUDIT

PSM or RMP audits have multiple objectives. Most important is the assurance an audit brings or the adequacy of implementation of the particular program. Specifically, the audit measures program effectiveness, identifies deficiencies, and verifies correction of previously identified deficiencies. The audit also provides management with a status assessment and may recommend improvements and identify good practices. The material in this section is based on a portion of the content of a Nuclear QA Certification Course taught at Sandia National Laboratories and several companies, the ASQ Nuclear Auditor Training Manual (ASQ 1986), the *System Safety Analysis Handbook* (Stephans and Talso 1997), and the OSHA Inspection Manual (U.S. Department of Labor 1994).

As an overview, the mechanics of an individual audit are shown in Figure 24-1. The figure depicts the general sequence and procedures used for the formal external audit by an independent review team, but it can also apply to

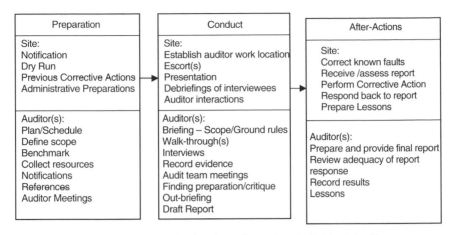

Preparation	Conduct	After-Actions
Site: Notification Dry Run Previous Corrective Actions Administrative Preparations	Site: Establish auditor work location Escort(s) Presentation Debriefings of interviewees Auditor interactions	Site: Correct known faults Receive /assess report Perform Corrective Action Respond back to report Prepare Lessons
Auditor(s): Plan/Schedule Define scope Benchmark Collect resources Notifications References Auditor Meetings	Auditor(s): Briefing – Scope/Ground rules Walk-through(s) Interviews Record evidence Audit team meetings Finding preparation/critique Out-briefing Draft Report	Auditor(s): Prepare and provide final report Review adequacy of report response Record results Lessons

Figure 24-1. Mechanics Overview of an Individual Audit

the small group or individual auditor. Tasks of both the auditors and site personnel are shown. Timeline for the audit may take a month overall with several weeks for preparation and one week each for conduct and after actions.

The conduct of the audit is shown in the center block of the figure and during this phase, it is mandatory that the auditor performs in a particularly professional manner. For OSHA inspectors and state plan states (where OSHA has approved the state administering their own occupational safety and health programs), there must be trained compliance safety and health officers with experience in the chemical industry that lead the PQV inspection. Keep in mind that the auditor's function is to assess to the established criteria and obtain objective evidence of program adequacy. It is not to provide individual advice or to recount prior experiences. It is necessary that a clear understanding of the audit purpose and a level of rapport between the site and the audit team should be established from the start. The auditors are at the site to "get the facts," and separate the trivial from the important. Auditors should begin in a friendly way and demonstrate an attitude that the auditor is on an equal basis with those he or she is auditing. Auditors should be empathetic and not sympathetic in that they understand the feelings, thoughts, and attitudes of the site personnel being interviewed, but not necessarily agreeing that they are correct. Sympathy is agreement. But it is important to understand the significance of actions that take place within the site. For example, as noted in the OSHA Inspection Manual, 11 PSM elements can be affected by changing one valve and it's important that the auditors confirm that required follow-ups are accomplished by the action on the part of site personnel.

When the report is drafted, written preparation of the individual findings is important. If only the OSHA PSM Checklist is used, the task is easier, but application to the EPA RMP and in the nuclear area it may be more difficult.

Generally, findings fall into two categories—those directly in contravention with a stated rule or standard requirement, and those that should be noted to improve operations or safety. Both need to be coordinated with and approved by the audit team leader before being disclosed to the audited site. Each finding needs to have back-up justification with evidence. The evidence should be in the form of site records or the record of the auditor's interview with one or more site personnel.

LESSONS

In part, continuous improvement can be achieved by learning and recording the lessons from those who have done similar work in the past. Note that many are repeats of previous lessons. They are listed because of their significance and the need not to repeat any correctible shortcomings of the past.

Having been on both sides of these inspections, there are some lessons that may appear to be fairly easily learned and implemented, but in the pressure of day-to-day activity and sometimes more work than one would like, it is possible to forget the essentials in preparing for administering and receiving an external or internal audit. Therefore, it is important to record some of the fundamental lessons so that the inspection or audit is conducted efficiently from the standpoint of both the inspector and the recipient of the inspection. The following are some of the more important lessons and by no means an exhaustive listing. It can be added to as your personal experience is gained.

Lessons from the Site and the Inspection Recipient

- Show you care and be hospitable. If possible, provide a private workspace, escorts for individual auditors, a set of references and site information, and a point of contact to provide other materials or respond to questions. Donuts for the inspection team each day would be appreciated. Also, as correctable deficiencies are discovered, they should be corrected on the spot. During an audit, a site might consider having a second shift of one or more persons reacting to and performing corrective action if needed.
- Function professionally. Provide a formal presentation about the site and processes. Respond to audit team needs in a timely fashion.
- Put your best foot forward. From the first time the auditors enter the site until they leave, they should be favorably impressed. Ensure that the site is clean and presentable and that your best people are the walk-through escorts and are the ones interviewed.
- Rehearse. If nothing else, walk through the areas to be visited by the auditors and provide guidance to those being interviewed regarding how to conduct themselves (and to report back).
- Communicate. Keep people informed and meet on a daily basis with those in contact with the auditors. Interact with the Inspecting team leader.

- Implement efficiently. When implementing process safety, do both 40 CFR 68 (RMP) and 10 CFR 1910.119 (PSM) in parallel as much as appropriate.
- Best surprise is no surprise. People don't like surprises and this includes auditors. As much as possible try to anticipate the audit team needs.
- Know what is expected. Be alert to communicate the audit team needs and reiterate the ground rules stated during the first day's audit team presentation.
- Ensure records and traceability. Factual evidence is based on records and other documents. Work authorization, maintenance forms, operating logs, and similar records should be available and accessible.

Lessons from the Auditor/Lead Inspector

- Know the requirements. Knowing the requirements and their interpretation is key to a successful audit. Every auditor must know the requirements.
- Do your homework. The preparation phase of the audit is most important to ensuring a successful audit. It is imperative that the lead auditor maps out each day and anticipates the issues that may arise.
- Use trained and qualified personnel. This almost goes without saying, but at times there is management pressure to either orient new personnel or to go with a less than adequate team. This becomes a challenge for the team lead. Personnel should be selected as auditors also based on their ability to interact with the personnel at the site.
- Review and emphasize the ground rules. It is vital that both the auditor and the site personnel understand the audit ground rules. This is accomplished at the in-briefing. At that time discussion of schedule, the interview process, records to be reviewed, and finding takes place.
- Communicate. Hold daily meetings after each day's inspection to review and prepare findings.
- Focus on factual, not subjective. In some instances, factual results can seem counterintuitive, but it is factual material that supports an adequate evaluation.
- When finished, leave. Too many times, the site has a tendency to argue the finding results if they are not favorable.

Chemical Safety and Hazard Investigation Board

The same legislation (Clean Air Act Amendments of 1990) that established standards for OSHA PSM and the EPA Risk Management Program established the Chemical Safety and Hazard Investigation Board (CSB). The CSB can be thought of as to the chemical process industry what the National Transportation Safety Board is to the aircraft and airline industry. More specifically, the mission of the U.S. Chemical Safety and Hazard Investigation Board is to promote the prevention of major chemical accidents at fixed facilities. To achieve the mission, the CSB established the following vision for the organization as it becomes fully operational:

- Produce an average of five accident investigation reports and one hazard investigation report each year.
- Have an effective recommendations program.
- Have a diverse, highly trained, productive workforce committed to continuous improvement and prevention of chemical accidents.
- Have an effective board providing strategic direction and oversight.
- Have a shared commitment with key stakeholders and customers on CSB mission, vision and goals.
- Have well-established public and private partnerships.

The CSB also conducts research, advises industry and labor on actions they should take to improve safety, and makes recommendations to local, federal, and state agencies such as the EPA and the OSHA at the U.S. Department of Labor, the key federal entities regulating industries that use chemicals. As set forth in the CSB's strategic plan, the target date to achieve full operational status is fiscal year 2005. Much more information about the CSB is available on its website at www.csb.gov.

SUMMARY

Both the PSM and RMP rules state that employers must certify that they have evaluated compliance to verify that the procedures and practices developed under the standard are adequate and are being followed. This chapter provided information about the relationship of audits to QA, the audits themselves, audit conduct, lessons learned, and some related information about the Chemical Safety and Hazard Investigation Board. It was stressed that the OSHA Inspection Manual is an essential reference at any site with process safety responsibilities.

A copy of the OSHA Inspection Manual is a valuable tool for the safety professional at the site where internal or external PSM/RMP inspections are conducted. The manual has a wealth of vital information.

REVIEW QUESTIONS

1. Why it is so important to review the OSHA Inspection Manual's checklists? (A "no" response is sufficient to award a violation that could be costly for the facility.)

2. What are the ties between quality assurance and process safety?
3. What are two major purposes for audits?
4. What are some of the differences between an internal and external audit?
5. Name four site actions to be taken during each of the three phases of an audit.
6. Name four audit team actions to be taken during each of the three phases of an audit.
7. Can the OSHA Inspection Manual also be used as reference for the Risk Management Program audit?
8. Name three lessons learned by recipients of PSM/RMP audits.
9. Name three lessons learned by auditors of PSM/RMP audits.
10. What are four of the functions of the Chemical Safety and Hazard Investigation Board?

REFERENCES

American Society for Quality (now ASQ), Energy Division, *Nuclear Quality Systems Auditor Training Handbook, 2nd ed.*, Milwaukee, WI, 1986.

ASME/ANSI, NQA-1, *Quality Assurance Requirements for Nuclear Facility Applications*. American Society of Mechanical Engineers, New York, NY, 2001.

Stephans R.A. and Talso W.W., eds., *System Safety Analysis Handbook*, 2nd ed. (Unionville, VA: System Safety Society, 1997).

U.S. Department of Labor, Occupational Safety and Health Administration, 29 CFR 1910.119, *Process Safety Management of Highly Hazardous Chemicals*, 1992.

U.S. Department of Labor, OSHA, Directorate of Compliance, CPL 2-2.45A CH-1, *29 CFR 1910.119, Process Safety Management of Highly Hazardous Chemicals—Compliance Guidelines and Enforcement Procedures*, September 13, 1994.

U.S. Environmental Protection Agency, 40 CFR 68, Risk Management Program, 1996.

PART VI

PROFESSIONALISM AND PROFESSIONAL DEVELOPMENT

CHAPTER 25

Professionalism and Professional Development

While most recent graduates and those who are just beginning in the safety field aren't necessarily thinking about professionalism or their responsibilities as a safety professional, they should. At least they should be aware of what this safety profession entails and what they should be looking forward to. Most importantly, safety practitioners need to understand that, as a professional, there is more responsibility required than receiving basic safety training and performing in a competent manner.

INTRODUCTION

The purpose of this portion of the book is to provide information related to professionalism as it relates to system safety; identify professional societies interested in system safety, discuss continuing education, and provide advice from someone who has been in this field and other fields for a number of years. The section is also a form of mentoring to those who are new to the system safety discipline or who are contemplating the safety (system safety) profession.

For more information on the safety profession as a career, it is suggested that the *Career Guide to the Safety Profession* be reviewed. This 50-page document provides the who, what, where, and why of the safety profession and a number of added sources of related information. It is a joint publication of the Board of Certified Safety Professionals and the American Society of Safety Engineers Foundation.

WHAT IS PROFESSIONALISM?

In this section, both dictionary definitions of professionalism and the interpretations of prominent system safety professionals in the field are presented.

System Safety for the 21ˢᵗ Century: The Updated and Revised Edition of System Safety 2000, by Richard A. Stephans
ISBN 0-471-44454-5   Copyright © 2004 John Wiley & Sons, Inc.

Merriam Webster (Merriam Webster 1993) defines a professional in part as characterized by or conforming to the technical or ethical standards of a profession. The *American Heritage Dictionary* (Morris 1981) defines professionalism as professional status, methods, character or standards.

According to Curt Lewis, System Safety Manager for American Airlines, "A system safety professional is someone who not only who knows the job, but who also has integrity and who can communicate the importance of safety to the company's mission to those above and below. The professional continuously seeks to improve his/her knowledge and ability" (C.E. Lewis, personal communication, August 13, 2003).

Perry D'Antonio, a manager at the Sandia National Laboratories and a former International President of the System Safety Society, says that "I view professionalism as transient in nature—much like an electrical circuit in which you need a continuous power source to ensure a fully charged circuit. In this metaphor, professional development is the power source. Professionalism is achieved when an individual's methods, behavior, skill, performance, and spirit of dedication to excel are commensurate with other recognized professional practitioners in the same occupation. Thus, the state of professionalism requires continuous dedication to appropriate professional development activities to ensure the professional practitioner maintains technical excellence in an ever changing world" (P.E. D'Antonio, personal communication, August 13, 2003).

Professionalism may also be thought of as a demeanor, that is, the way someone conducts himself or herself. Attributes such as a person's presence, succinct speech, and even dress contribute to a professional persona. A professional is someone that others rely on for knowledge in a particular area and a person who is regarded as an expert.

What Is a Safety Professional?

As stated by the Board of Certified Safety Professionals (BCSP 2003),

> "a safety professional is a person engaged in the prevention of accidents, incidents, and events that harm people, property, or the environment. They use qualitative and quantitative analysis of simple and complex products, systems, operations, and activities to identify hazards. They evaluate the hazards to identify what events can occur and the likelihood of occurrence, severity of results, risk (a combination of probability and severity), and cost. They identify what controls are appropriate and their cost and effectiveness. Safety professionals make recommendations to managers, designers, employers, government agencies, and others. Controls may involve administrative controls (such as plans, policies, procedures, training, etc.) and engineering controls (such as safety features and systems, fail-safe features, barriers, and other forms of protection). Safety professionals may manage and implement controls.

> Beside knowledge of a wide range of hazards, controls, and safety assessment methods, safety professionals must have knowledge of physical, chemical, bio-

logical and behavioral sciences, mathematics, business, training and educational techniques, engineering concepts, and particular kinds of operations (construction, manufacturing, transportation, etc.)"

More to the point, your question might be "what are the characteristics of a system safety professional? In this author's opinion, a system safety professional needs to go beyond the narrow definition of a safety professional as stated above and he or she should have the following attributes:

1. Possess basic education and skills. Many of us come into the safety field from others and learn as we go, but some basic skills are essential. If you come to the job without formal system safety training, get it as soon as you can.
2. Exhibit and maintain personal integrity. It may be a cliché, but the points of the Boy Scout Law learned early in my youth have remained with me throughout my career: trustworthy, loyal, helpful, friendly, courteous, kind, obedient, cheerful, thrifty, brave, clean, and reverent. They are valid today and will be in the future.
3. Focus on the safety mission. With a great deal of work every day and in many areas, it's easy to become side-tracked. Don't let that happen.
4. Seek continuing education. Professional education is not a one-time proposition! It necessarily is continuous. One way to "force" this need for continuing education is through certification in the field. Because after one is certified by a reputable organization, there are requirements for "continuance of certification" that entail periodic training and related activities to maintain an appropriate skill level. You will also learn from your professional peers.
5. Contribute to the profession. Start early by joining a professional society (or more than just one). As you mature in the profession, advance in the society and assist others who are new.

With this advice from an old-timer in the field, it's appropriate to reinforce what was said with information from the By-Laws Principles and Canons of the System Safety Society. The complete text is also provided in the Appendix. The following is an excerpt from those fundamental principles and canons of the System Safety Society (C.E. Lewis, personal communications, August 13, 2003 and appendix):

"System Safety professionals shall undertake, and encourage their employees to further their education, become registered and certified by appropriate legal and professional authorities at the earliest possible date, to join and participate in appropriate professional organizations and to attend and present papers at professional and technical Society meetings.

PROFESSIONAL DEVELOPMENT

The objective of professional development is to ensure competence by means of continuing education and experience throughout one's career. Professional development becomes increasingly important as the professional continues in the field over the years. In a field where there are technological advances, the professional in that field needs to keep up with those advances in order to retain his or her status as a professional. In some fields this is mandated and in others the requirement may be more subtle: The work and business context continually changes and requires continuous learning in order to be effective. The professional's toolbox changes and one must keep an individual's personal toolbox up to date. This can be accomplished after basic and specialty education with continuing education, experience, and advancement in the field.

A means of allowing for both professional development that leads to advancement in the field is to seek certification as a safety professional (Table 25-1). In a way, safety certification tends to provide the structure and impetus for achieving professional development. If certification is desired, one should certainly investigate the most efficient way to achieve a certification. There are many certifications and designations where you simply send money and receive a certificate! But having this type of designation may more likely have a negative affect. Other professionals (who may be looking at that same resume) tend to know which certifications are worth something and which have little meaning.

ACCREDITATION OF CERTIFICATIONS

By definition, accreditation in the safety (certification) field is the process of independently assuring a certification meets a minimum set of national standards. The standards go beyond technical requirements and ensure that the examination is fair, properly developed, and administrated. Tables 25-2 and 25-3 provide information about accrediting organizations and the approximate number of safety and health, environmental, and ergonomics certification programs that are accredited.

TABLE 25-1. Safety, Health, Environmental and Related Certifications*

Safety and Health	Environmental	Ergonomics
~25	~150	~15

*Figures are estimates from literature research over several years by the BCSP staff and reflect approximate/known numbers.

TABLE 25-2. U.S. Certification Accrediting Organizations

Name	Current Number of Organizations with Certifications Accredited*	Comments
National Commission for Certifying Agencies (NCCA)	40	Established in 1980 Primarily accredits certifications in the medical and related fields
Council of Engineering and Scientific Specialty Boards (CESB)	7	Established in 1991
National Skill Standards Board (NSSB)	6	NSSB "recognizes" programs as meeting programs meeting their excellence standards. It is a U.S. Federal Agency operating with Departments of Labor and Commerce and has been operating since 2001. Fifteen of the applicants failed to achieve recognition by 2003.
International Standards Organization—ISO (ANSI/ISO in U.S.)	6 pending	Six pilot certification organizations selected in August 2002 and began the process under ISO 17024.

*Accreditation may be by individual certification or granting organization. In the case of NSSB, recognition is granted.

TABLE 25-3. Accredited Safety, Health, Environmental and Related Certifications*

Safety and Health	Environmental	Ergonomics
~4	~10	~2

*Figures are estimates from literature research over several years by the BCSP staff and reflect approximate/known numbers.

Future Direction of Safety and System Safety Professionalism

There are a number of potential future directions for the safety field and for system safety professionals. First, there are the international pressures of the global market and "international" standards that may apply to professionals.

Generally, there will be better educated professionals in the field. Better decisions will be made and a safer outcome will occur. That's the ultimate reason for us being in the field and it's gratifying. Also, there should be more competition between professionals and that will result in a higher standard for safety professionals. One has to realize that with a higher standard for professionals, it means that the individual must seek the higher levels of achievement. For example, in the past, a bachelor's degree was not necessarily a requirement for professionals in the safety field, now it is increasingly more important. In fact, to achieve professional certification a degree is often required, where 20 years ago, it was not.

Overall, there will be more safety information available and also, added tools in order to obtain that information. We will continue to be deluged and that is a blessing as well as a challenge.

The recent trend has been toward standardizing the number of courses taken for a bachelor degree and reducing that for a U.S. engineering degree— 130 credits for graduation (versus 160 several decades ago).

WHY BECOME CERTIFIED?

An accredited and recognized safety certification (whether CSP, CIH, or CHP) has value to you and others in several levels:

Personal. In addition to an enhanced self-image, the certification "made me learn the subject matter." Our individual jobs tend to concentrate in one area—the safety certification forces one to take a broader safety perspective and know more about the field. Most definitely and proven by survey, an average increased income of $15K per year throughout a career (in year 2000 dollars).

Peers. In the eyes of your peers, achievement of safety certification demonstrates that you have made the effort to "prove" your ability to meet a high professional standard in the field. In the System Safety Society, it provides higher point-value when advancing toward the achievement of "Fellow" designation within the Society.

Employer. The certification keeps your employer ahead by the ability to compete and win contracts and it moves the holder of the certification ahead in selection for jobs, bonuses and promotion. There is also a measure of the value of certification based on inclusion in government regulations, title protection legislation, government and commercial contracts, and in employer policies.

Public. Perhaps the greatest value is to the public in terms of the reliability of having a tested and proven competent person performing safety duties and the credibility that it brings.

Continuance of certification requirements ensures that you achieve the education or equivalency to keep you current in the field.

What's Best for Me?

You are (or are about to be) a professional. You need to spend some time and thought and make an assessment of the value and your options related to professional certification and or licensure.

Cost/Benefit Analysis—Cost in terms of $, rigor (usually two exams), time (probably takes much more than a year).

Benefit—$ ($300K in a career for the top safety certification based on the $15K per year amount), forced professional development, bragging rights.

Which Certification?

How do I choose from the 225 or more certifications/designations available? There are several potential ways. One way of choosing might be by identifying the number if accreditations achieved by the certifying organization. Another is by looking at "want ads." Find out what employers want. Another is by word of mouth in your circle of professionals. Regarding number of accreditations, Table 25-4 is of interest.

Although it is possible to pay some money and get some form of recognition without any examination of your skill, this may be look at by peers and employers as a negative and not in your favor. So, carefully consider the lesser thought of designations and certifications. But again, it's up to you.

More than One Certification?

One advantage is that the initial examination may be waived if you hold a current certification as a PE, CIH, or CSP. Another advantage is that Contin-

TABLE 25-4. Numbers of Accreditations

Safety Certification	Accreditation or Equivalent*
Certified Safety Professional	4
Certified Industrial Hygienist	1
Certified Health Physicist	1
Others (Safety Specialty under CCHEST)	1

*Equivalent being a Federal Government approval of the certification granting process.

uance of Certification points may be applied to both (or more) certifications/licenses. The most important reason is that the multiple certifications tend to make one a more versatile individual and capable of performing a variety of tasks. With this is also the potential for additional salary. Quality certifications also add additional pay value when two or more are held.

Example of the Time it Takes—CSP

Unless you are very smart and experienced in many facets of the safety certification you are pursuing, you should expect a period of more than one year from the time of making application until achieving certification. The process does take time and a *commitment* on the part of the person seeking certification. Commitment is underscored because there must be a strong desire to not just achieve the certification, but also to follow through on the individual steps along the way. Just the application is not a simple task. It takes administrative information, submission of college transcripts, detailing professional experience history, contacting references, and having the package notarized.

Based on information from the Board of Certified Safety Professionals, Table 25-5 shows the timeline experience with all candidates from the time of application until award of the CSP. Also, it is interesting to note that the average new CSP was approximately 39 years of age. The youngest to achieve certification was 24 and the oldest was 74 years old.

As you might expect, in the real world, one does not usually get ahead without working for it. By the same token, there are many choices when it comes to professional development. One can choose to take the path of least resistance and "purchase" some form of certification or registration that does not entail a demonstration of comprehensive knowledge in the profession. However, this can also identify you as a person who is willing to slide by and it may actually work against you.

So, it is up to you. You may choose to do nothing, wait, or do it now. You can imagine my recommendation.

Advice

Growing up, I was frequently reminded about the quotation from Thomas Edison, the prolific inventor, that success was 10% inspiration and 90% perspiration. If you were my son or daughter and you were new (or old) to the profession, I would advise you as follows:

TABLE 25-5. Time to Achieve CSP from the Time of Application

Average:	*3.5 years*
Median:	*3.3 years*
Shortest Time:	*46 days*
Longest Time:	*8 years*

- Don't wildly jump into a field and move on to something else just because you were not immediately successful. It takes time to acquire knowledge and experience (and success).
- Become somewhat "thick-skinned." Often as a safety specialist, you are the bearer of bad news, and keep in mind that there is always the tendency to "kill the messenger." That's a natural tendency, so be ready to cope with it.
- Continue your education! Get a masters degree or higher. Take advantage of company-sponsored training as it is presented. Part of that is to become certified in your specialty as soon as you can. Many leadership positions require advanced degrees.
- Take the examinations for professional certification as soon after school as you can after college. The earlier you acquire academic and certification credentials, the earlier you benefit from the additional pay. On the other hand, the longer it has been since your academic training, the more difficult the process. But it is still possible to achieve certification later in life (mine was granted by examination after the age of 60).
- Don't bite off too big a chunk. And if you do, break the job into manageable pieces that can be attacked one at a time.
- Become active in a professional organization. It can only help in a variety of ways—technical, leadership, recognition, and giving back to your profession and learning from other professionals.
- Don't necessarily take my advice. Seek information from others as well. Then make a decision and follow through. Finishing what you start is often the most difficult part.

SUMMARY

Professional development and keeping current in your field is the responsibility of every professional and it includes continuing education and maintaining competence in the field. Accredited safety certification will help to assure that you are kept up to date and continue your competency.

REVIEW QUESTIONS

1. Which safety certification has the greatest number of national accreditations?
2. What is the difference in jurisdiction between certification and licensure?
3. Name three national and internationally recognized accrediting bodies for certifications.
4. What is the objective of professional development?
5. Who has responsibility for your professional development?

6. Name three accredited safety or related certifications.
7. What are some of the attributes of a safety professional?
8. What are four functions of a professional safety position as defined by the ASSE?
9. According to the System Safety Society principles and cannons, what are some of the requirements of a system safety professional?
10. Why should one consider becoming certified as a safety professional?

REFERENCES

Board of Certified Safety Professionals Website, available at www.bcsp.org. Accessed May 2003.

Merriam Webster Collegiate Dictionary, 10th ed. Merriam Webster, Springfield, MA, 1993.

Morris W., ed., *The American Heritage Dictionary of the English Language.* Houghton Mifflin Company, Boston, MA, 1981.

System Safety Society Website, Society Constitution & By-Laws available at www.systemsafety.org. Accessed May 2003.

The Scope and Functions of the Professional Safety Position*

Functions of the Professional Safety Position

The major areas relating to the protection of people, property and the environment are:

A. Anticipate, identify and evaluate hazardous conditions and practices.

B. Develop hazard control designs, methods, procedures and programs.

C. Implement, administer and advise others on hazard controls and hazard control programs.

D. Measure, audit and evaluate the effectiveness of hazard controls and hazard control programs.

A. Anticipate, identify and evaluate hazardous conditions and practices.

This function involves:

1. Developing methods for
   a. anticipating and predicting hazards from experience, historical data and other information sources.
   b. identifying and recognizing hazards in existing or future systems, equipment, products, software, facilities, processes, operations and procedures during their expected life.
   c. evaluating and assessing the probability and severity of loss events and accidents which may result from actual or potential hazards.
2. Applying these methods and conducting hazard analyses and interpreting results.

* Published by the American Society of Safety Engineers, 1993 Revision 1800 East Oakton Street, Des Plaines, IL 60018-2187, with permission and appreciation.

System Safety for the 21st Century: The Updated and Revised Edition of System Safety 2000, by Richard A. Stephans
ISBN 0-471-44454-5   Copyright © 2004 John Wiley & Sons, Inc.

3. Reviewing, with the assistance of specialists where needed, entire systems, processes, and operations for failure modes, causes and effects of the entire system, process or operation and any sub-systems or components due to
   a. system, sub-system, or component failures.
   b. human error.
   c. incomplete or faulty decision making, judgements or administrative actions.
   d. weaknesses in proposed or existing policies, directives, objectives or practices.
4. Reviewing, compiling, analyzing and interpreting data from accident and loss event reports and other sources regarding injuries, illnesses, property damage, environmental effects or public impacts to
   a. identify causes, trends and relationships.
   b. ensure completeness, accuracy and validity of required information.
   c. evaluate the effectiveness of classification schemes and data collection methods.
   d. initiate investigations.
5. Providing advice and counsel about compliance with safety, health and environmental laws, codes, regulations and standards.
6. Conducting research studies of existing or potential safety and health problems and issues.
7. Determining the need for surveys and appraisals that help identify conditions or practices affecting safety and health, including those which require the services of specialists such as physicians, health physicists, industrial hygienists, fire protection engineers, design and process engineers, ergonomists, risk managers, environmental professionals, psychologists, and others.
8. Assessing environments, tasks and other elements to ensure that physiological and psychological capabilities, capacities and limits of humans are not exceeded.

B. Develop hazard control designs, methods, procedures and programs.

This function involves:

1. Formulating and prescribing engineering or administrative controls, preferably before exposures, accidents, and loss events occur, to
   a. eliminate hazards and causes of exposures, accidents and loss events.
   b. reduce the probability or severity of injuries, illnesses, losses or environmental damage from potential exposures, accidents, and loss events when hazards cannot be eliminated.
2. Developing methods which integrate safety performance into the goals, operations and productivity of organizations and their management and into systems, processes, and operations or their components.

3. Developing safety, health and environmental policies, procedures, codes and standards for integration into operational policies of organizations, unit operations, purchasing and contracting.
4. Consulting with and advising individuals and participating on teams
   a. engaged in planning, design, development and installation or implementation of systems or programs involving hazard controls.
   b. engaged in planning, design, development, fabrication, testing, packaging and distribution of products or services regarding safety requirements and application of safety principles which will maximize product safety.
5. Advising and assisting human resources specialists when applying hazard analysis results or dealing with the capabilities and limitations or personnel.
6. Staying current with technological developments, laws, regulations, standards, codes, products, methods, and practices related to hazard controls.

C. Implementation, administer and advise others on hazard controls and hazard control programs.

This function involves:

1. Preparing reports which communicate valid and comprehensive recommendations for hazard controls which are based on analysis and interpretation of accident, exposure, loss event and other data.
2. Using written and graphic materials, presentations and other communication media to recommend hazard controls and hazard control policies, procedures and programs to decision-making personnel.
3. Directing or assisting in planning and developing educational and training materials or courses. Conducting or assisting with courses related to designs, policies, procedures and programs involving hazard recognition and control.
4. Advising others about hazards, hazard controls, relative risk and related safety matters when they are communicating with the media, community and public.
5. Managing and implementing hazard controls and hazard control programs which are within the duties of the individual's professional safety position.

D. Measure, audit and evaluate the effectiveness of hazard controls and hazard control programs.

This function involves:

1. Establishing and implementing techniques which involve risk analysis, cost, cost-benefit analysis, work sampling, loss rate and similar method-

ologies, for periodic and systematic evaluation of hazard control and hazard control program effectiveness.

2. Developing methods to evaluate the costs and effectiveness of hazard controls and programs and measure the contribution of components of systems, organizations, processes and operations toward the overall effectiveness.

3. Providing results of evaluation assessments, including recommended adjustments and changes to hazard controls or hazard control programs, to individuals or organizations responsible for their management and implementation.

4. Directing, developing, or helping to develop management accountability and audit programs which assess safety performance of entire systems, organizations, processes and operations or their components and involve both deterrents and incentives.

The Scope of the Professional Safety Position
The perform their professional functions, safety professionals must have education, training and experience in a common body of knowledge. Safety professionals need to have a fundamental knowledge of physics, chemistry, biology, physiology, statistics, mathematics, computer science, engineering mechanics, industrial processes, business, communication and psychology. Professional safety studies include industrial hygiene and toxicology, design of engineering hazard controls, fire protection, ergonomics, system and process safety, safety and healthy program management, accident investigation and analysis, product safety, construction safety, education and training methods, measurement of safety performance, human behavior, environmental safety and health, and safety, health, and environmental laws, regulations and standards. Many safety professionals have backgrounds or advanced study in other disciplines, such as management and business administration, engineering, education, physical and social sciences and other fields. Others have advanced study in safety. This extends their expertise beyond the basics of the safety profession.

Because safety is an element of all human endeavors, safety professional perform their functions in a variety of context in both public and private sectors, often employing specialized knowledge and skills. Typical settings are manufacturing, insurance, risk management, government, education, consulting, construction, health care, engineering and design, waste management, petroleum, facilities management, retail, transportation, and utilities. Within these contexts, safety professionals must adapt their functions to fit the mission, operations and climate of their employer.

Not only must safety professionals acquire the knowledge and skill to perform their functions effectively in their employment context, through continuing education and training they stay current with new technologies, changes in laws and regulations, and changes in the workforce, workplace and world business, political and social climate.

As part of their positions, safety professionals must plan for and manage resources and funds related to their functions. They may be responsible for supervising a diverse staff of professionals.

By acquiring the knowledge and skills of the profession, developing the mind set and wisdom to act responsibly in the employment context, and keeping up with changes that affect the safety profession, the safety professional is able to perform required safety professional functions with confidence, competence and respected authority.

System Safety Society Fundamental Principles and Canons

SECTION 1—Fundamental Principles
System Safety professionals shall uphold and advance the integrity, honor and stature of the System Safety profession by:

1.1 Applying acquired knowledge and skills for the benefit and welfare of mankind.

1.2 Assuring the best interests of the public, employer, client and customer are objectively established and maintained in all professional activities.

1.3 Striving to increase the competence and effectiveness of the System Safety profession.

1.4 Supporting and participating in the activities of professional safety and related organizations.

SECTION 2—Fundamental Canons

2.1 System Safety professionals shall hold paramount the safety, health and welfare of mankind in the performance of their professional duties.

2.2 System Safety professionals shall perform services only in the areas of their competency.

2.3 System Safety professionals shall be objective and truthful in all of their professional actions, documents and statements.

2.4 System Safety professionals shall perform professional services as faithful agents and trustees for each employer, client or customer, void of any known conflict of interest.

2.5 System safety professionals shall build their professional reputations on the merits of their services; void of deceitful and unethical practices.

System Safety for the 21ˢᵗ Century: The Updated and Revised Edition of System Safety 2000, by Richard A. Stephans
ISBN 0-471-44454-5   Copyright © 2004 John Wiley & Sons, Inc.

2.6 System Safety professionals shall associate only with reputable persons and organizations.

2.7 System Safety professionals shall continue their professional development throughout their careers and shall provide opportunities for the professional development of persons under their supervision.

ARTICLE IV GUIDELINES FOR USE WITH THE FUNDAMENTAL CANONS OF ETHICS

SECTION 1—System Safety professional Shall Hold Paramount the Safety, Health, and Welfare of Mankind in the Performance of Their Professional Duties.

1.1 System Safety professionals must recognize that the safety, health, and welfare of mankind is dependent upon sound scientific, engineering and management principles and practices being incorporated into the design, test, production, construction, and use of products and systems.

1.2 System Safety professionals shall approve only plans, specifications, designs, hardware and software that are safe and in conformity with accepted safety standards.

1.3 System Safety professionals shall assure the development of standards, codes, procedures or other similar documentation to enable the public and clients to understand the degree of safety, risk and life expectancy associated with the use and operation of the products and systems for which they are responsible.

1.4 System Safety professionals shall inform their client or employer and, as appropriate, proper authorities of the possible consequences whenever their professional judgment is overruled under circumstances where the safety, health, and welfare of the public is endangered.

1.5 System Safety professionals who know, or believe, another person or firm is in violation of any of the provisions of these Guidelines, shall report such information to that person or firm and, as appropriate, to the proper authorities.

1.6 System Safety professionals shall accept opportunities to enhance the safety, health and welfare of their communities.

SECTION 2—System Safety Professionals Perform Services only in the Areas of their Competency.

2.1 System Safety professionals shall assume only safety, task, for which they are qualified by education or experience in the specific field of concern.

2.2 System Safety professionals shall affix their signatures and seals only to documents dealing with subject matter in which they have competency by virtue of education or experience and only to safety plans or documents prepared under their direct supervisory control.

SECTION 3—System Safety Professionals are objective and truthful in all of their professional documents and statements.

3.1 System Safety professionals shall endeavor to extend public knowledge and understanding of the achievements, capabilities and limitations of engineering technology.

3.2 System Safety professionals shall include all relevant and pertinent facts and information in their professional reports, statements, or testimony.

3.3 System Safety professionals, when serving as part expert or technical witness before any court-commission, or other tribunal, shall express opinions only when they are founded upon adequate knowledge of the facts at issue, from a background of technical competence in the subject matter, and upon honest conviction of the accuracy and propriety of their testimony.

3.4 System Safety professionals shall not issue statements criticisms or arguments on safety matters which are inspired or paid for by interested individuals or entities unless they preface their comments by explicitly disclosing the identities and interests of the parties on whose behalf they are speaking, and by revealing the existence of any financial or other interest they have in the matters involved.

3.5 System Safety professionals shall be dignified and modest in describing their work and merit, and shall avoid any statement or act tending to promote their own interests at the expense of the integrity, honor and dignity of the profession.

SECTION 4—System Safety Professionals Perform Professional Services as Faithful Agents and Trustees for Each Employer, Client or Customer, Void of any Known Conflict of Interest.

4.1 System Safety professionals shall avoid known conflicts of interest with their employers, clients or customers, and promptly inform such persons of any business association, interest, or circumstances which could influence their judgment or the quality of their services.

4.2 System Safety professionals shall not accept compensation, financial or otherwise, from more than one party for the exact same service on the same project, unless the circumstances are fully disclosed and agreed to by all interested parties.

4.3 System Safety professionals shall treat information acquired in the course of their assignments as confidential, and shall not use such information as a means of enhancing personal profit or prestige if such action would be adverse to the interests of their clients, employers, customers, or the public.

4.4 System Safety professionals shall not accept professional employment outside of their regular place of employment without the knowledge of their employer.

SECTION 5—System Safety Professionals Shall Build Their Professional Reputations of the Merit of Their Services; Void of Deceitful and Unethical Practices.

5.1 System Safety professionals shall not pay or offer to pay (directly or indirectly) any commission, contribution, gift or other consideration in order to secure work, except for employment agency fees.

5.2 System Safety professionals shall negotiate contracts for professional services only when they are competent and qualified for the type of professional service required.

5.3 System Safety professionals shall negotiate and fully disclose an acceptable and rate of compensation with a client, employee or other professional commensurate with the scope of services to be performed.

5.4 System Safety professionals shall assure that representations of academic, experience and professional qualifications are factual for themselves and their associates. They do not misrepresent or exaggerate their degree of responsibility in, or the subject matter of, prior assignments.

5.5 System Safety professionals shall evaluate and provide reports to others on the work of other safety professionals only with the knowledge of such professionals unless the assignments or contractual agreements for the work by the other safety professional has been terminated or unless the functions and responsibilities of their position are specifically identified or implied to perform work evaluation.

5.6 System Safety professionals shall advertise only in dignified business and professional publications. They shall assure the advertisement is factual and not misleading with respect to their accomplishments or extent of participation in any services or projects described.

5.7 System Safety professionals shall prepare articles for the lay or technical press which are factual, dignified and free from self-serving ostentations or laudatory implications. Such articles shall factually depict their direct participation in the work described and give credit to others for their share of the work.

5.8 System Safety professionals shall extend permission for their name to be used in commercial advertisements (such as may be published by manufacturers, contractors, material suppliers, etc.) only by means of a modest dignified notation acknowledging their scope of participation in the project or product described.

5.9 System Safety professionals shall attempt to attract employees, from another employer by ethical and honorable means only.

5.10 System Safety professionals shall advertise for recruitment of personnel in appropriate publications or by special distribution, assuring the information presented is displayed in a dignified manner; restricted to firm name, address, telephone number, appropriate symbol, the fields of practice in which the firm is involved and factual descriptions of positions, available, qualifications required and benefits provided.

5.11 System Safety professionals shall not maliciously or falsely (directly or indirectly) injure the professional reputation prospects, practice or employment of another safety professional, nor indiscriminately criticize another's work.

5.12 System Safety professionals may, without compensation, perform professional services which are advisory in nature for civic, charitable, religious or nonprofit organizations.

5.13 System Safety professionals shall not use equipment, supplies, laboratory or office facilities of their employers to perform outside private practice without consent.

SECTION 6—System Safety Professionals Shall Associate only with Reputable Persons or Organizations.

6.1 System Safety professionals shall not knowingly associate with or permit the use of their name or firm name in business ventures or practices by any person or firm they know, or have reason to believe, is engaged in personal, business, or professional practice of a fraudulent, illegal or dishonest nature.

SECTION 7—System Safety Professionals Shall Continue Their Professional Development Throughout Their Careers and Provide Opportunities of the Professional Development of Persons Under Their Supervision.

7.1 System Safety professionals shall undertake, and encourage their employees to further their education, become registered and certified by appropriate legal and professional authorities at the earliest possible date, to join and participate in appropriate professional organizations and to attend and present papers at professional and technical Society meetings.

7.2 System Safety professionals shall give proper recognition and credit to others for professional development accomplishments and safety work performed, and recognize the proprietary interest of others. They shall identify the person or persons responsible for designs, inventions, writings or other accomplishments as appropriate.

Professional System Safety and Related Societies and Organizations

System Safety Society
www.system-safety.org
P.O. Box 70
Unionville, VA 22567-0070
800-747-5744
540-854-8630 (fax)
syssafe@ns.gemlink.com (e-mail)

The System Safety Society was organized in 1962 and incorporated in 1973 as an entity that promoted education and enhancement of the system safety discipline. System safety is an engineering and management discipline that emphasizes preventing accidents rather than reacting to them. In general, it does so by identifying, evaluating, analyzing, and controlling hazards throughout the life cycle of a system. The discipline is an outgrowth of the aerospace industry and was first applied to missile systems and to the Apollo Program to land a man on the moon. Currently, system safety practitioners are contributing to safety in industry, government, and academia throughout the world.

The Society's objectives are to:

- Advance the System Safety state-of-the-art.
- Contribute to the understanding of System Safety and its applications.
- Disseminate newly developed knowledge about System Safety.
- Further professional development.
- Improve public perception about hazards and the System Safety discipline.
- Improve communication between the System Safety discipline and other professional groups.

System Safety for the 21ˢᵗ Century: The Updated and Revised Edition of System Safety 2000, by Richard A. Stephans
ISBN 0-471-44454-5   Copyright © 2004 John Wiley & Sons, Inc.

- Establish standards for the System Safety discipline.
- Assist federal, state, and local government bodies concerned with safety.
- Establish standards for System Safety educational programs.

The System Safety Society publishes the *Journal for System Safety* and texts, which include the *System Safety Analysis Handbook* that has been sold in 38 countries. The Society also honors its own members annually for various achievements and provides a vast networking opportunity for members to solve problems and identify job opportunities.

The Society is a worldwide organization with members from more than 20 countries outside the United States. Each year the Society sponsors an International System Safety Conference and exhibition, which is attended by hundreds of safety professionals and others.

American Industrial Hygiene Association
www.aiha.org
2700 Prosperity Ave., Suite 250
Fairfax, VA 22031
703-849-8888

AIHA is the essential resource for information on occupational and environmental health and safety issues. Founded in 1939, AIHA is an organization of more than 11,500 professional members dedicated to the anticipation, recognition, evaluation and control of environmental factors arising in or from the workplace that may result in injury, illness, impairment, or affect the well being of workers and members of the community. The AIHA overseas and conducts the CIH.

American Society of Safety Engineers—Engineering Division
www.asse.org
1800 East Oakton St.
Des Plaines, IL 60018-2178
847-699-2929

Founded in 1911, ASSE is the world's oldest and largest professional safety organization. Its 33,000 members manage, supervise, and consult on safety, health, and environmental issues in industry, insurance, government, and education. ASSE has 12 divisions and 148 chapters in the U.S. and abroad. It also has 64 student sections. The engineering Division is most closely related to System Safety.

National Safety Council
www.nsc.org
1121 Spring Lake Dr.
Itasca, IL 60143-3201

The National Safety Council has served as the premier source of safety and health information in the United States since 1913. Along with national responsibilities, the Council carries out its mission on the community level through a network of more than 60 local chapters.

Society of Fire Protection Engineers
www.sfpe.org
7315 Wisconsin Ave.
Suite 1225W
Bethesda, MD 20814
303-718-2910

The Society of Fire Protection Engineers was established in 1950 and incorporated in 1971. It is the professional society representing those practicing in the field of fire protection engineering. The Society has approximately 4,000 members in the United States and abroad, 51 regional chapters, 10 of which are outside the United States.

American Institute of Chemical Engineers
www.aiche.org
3 Park Ave.
New York, NY 10016-5991

The American Institute of Chemical Engineers, AIChE, was founded in 1908. AIChE is a professional association of more than 50,000 members that provides leadership in advancing the chemical engineering profession. Its members are creative problem-solvers who use their scientific and technical knowledge to develop processes and design and operate plants to make useful products at a reasonable cost.

American Academy of Environmental Engineers
www.enviro-engrs.org
130 Holiday Court, Suite 100
Annapolis, MD 21401
410-266-3311

The American Academy of Environmental Engineers is dedicated to excellence in the practice of environmental engineering to ensure the public health, safety, and welfare to enable humankind to co-exist in harmony with nature. The Academy is dedicated to improving the practice, elevating the standards and advancing the cause of environmental engineering.

European Agency for Safety and Health at Work
http://agency.osha.eu.int
Some of the Focal Points are:

Arbetsmiljöverket
SE-171 84 Solna
Sweden
Contact person: Ms Elisabet Delang
+4687309119 (fax)
arbetsmiljoverket@av.se (e-mail)

Health and Safety Executive
Daniel House
Stanley Precinct
Bootle L20 3TW
United Kingdom
Contact person: Nicola Dorrian (neé Westhead)
0151 951 3191 (fax)
uk.focalpoint@hse.gsi.gov.uk (e-mail)

The Agency's principal safety and health information network is made up of a "Focal Point" in each European Union Member State, in the four EFTA countries and in the 13 candidate countries (CCs) to the European Union. This network is an integral part of the Agency's organization and plays an important role within the Agency structure. Focal Points are nominated by each government as the Agency's official representative in that country and are normally the competent national authority for safety and health at work.

The European Agency for Safety and Health at Work aims to make Europe's workplaces safer, healthier and more productive. It acts as a catalyst for developing, collecting, analyzing and disseminating information that improves the state of occupational safety and health in Europe. The Agency is a tripartite European Union organization and brings together representatives from three key decision-making groups in each of the EU's member states—governments, employers and workers' organizations.

Canada Safety Council
www.safety-council.org
1020 Thomas Spratt Place
Ottawa, ON K1G 5L5
Canada
613-739-1535
613-739-1566 (fax)

The Canada Safety Council is a national, nongovernment, charitable organization dedicated to safety. Our mission is to lead in the national effort to reduce preventable deaths, injuries and economic loss in public and private places throughout Canada. We serve as a credible, reliable resource for safety information, education and awareness in all aspects of Canadian life—in traffic, at home, at work and at leisure.

Focusing on safety education as the key to long range reduction in avoidable deaths and injuries, the Council serves as a national resource for safety programs, working with and through partner organizations who deliver and/or fund these programs.

Health Physics Society
www.hps.org
1313 Dolley Madison Blvd.
Suite 402
McLean, VA 22101
703-790-1745
703-790-2672 (fax)
hps@BurkInc.com (e-mail)

The Health Physics Society is a nonprofit scientific professional organization whose mission is to promote the practice of radiation safety. Since its formation in 1956, the Society has grown to approximately 6,000 scientists, physicians, engineers, lawyers, and other professionals representing academia, industry, government, national laboratories, the department of defense, and other organizations. Society activities include encouraging research in radiation science, developing standards, and disseminating radiation safety information. Society members are involved in understanding, evaluating, and controlling the potential risks from radiation relative to the benefits.

Board of Certified Safety Professionals (BCSP)
www.bcsp.org
208 Burwash Ave.
Savoy, IL 61874
217-359-9263

The Board of Certified Safety Professionals is a not-for-profit organization established in 1969 and devoted solely to setting standards for safety professionals and evaluating individuals against those standards. The BCSP's main certification is the Certified Safety Professional (CSP) designation. Currently, 10,000 individuals hold the CSP and about 5,000 are in the process of achieving the CSP. BCSP is independent of, but sponsored by the following societies and organizations:

- American Society of Safety Engineers
- American Industrial Hygiene Association
- System Safety Society
- Society of Fire Protection Engineers
- National Safety Council
- Institute of Industrial Engineers

Sources

ASSE: American Society of Safety Engineers Dictionary of Terms Used in the Safety Profession

SSDC: System Safety Development Center Glossary of SSDC Terms and Acronyms, SSDC-28 (DOE)

AFR 800-16: U.S. Air Force Regulation 800-16, USAF System Safety Programs (USAF)

MIL-STD-882B: Military Standard 882B, System Safety Program Requirements (DOD)

NSTS 22254: National Space Transportation System, Methodology for Conduct of NSTS Hazard Analyses (NASA)

Acceptable risk   The residual risk remaining after controls have been applied to associated hazards that have been identified, quantified to the maximum extent practicable, analyzed, communicated to the proper level of management and accepted after proper evaluation (SSDC).

Accident   An unwanted transfer of energy or an environmental condition which, due to the absence or failure of barriers and/or controls, produces injury to persons, property, or process (SSDC); as defined in NHB 5300.41(1D-2), "An unplanned event which results in an unsafe situation or operational mode" (NSTS 22254); an unplanned and sometimes injurious or damaging event which interrupts the normal progress of an activity and is invariably preceded by an unsafe act or unsafe condition or some combination thereof. An accident may be seen as resulting from a failure to identify a hazard or from some inadequacy in an existing system of hazard controls (ASSE).

Assumed risk   A specific, analyzed residual risk accepted at an appropriate level of management. Ideally, the risk has had analysis of alternatives for increasing control and evaluation of significance of consequences (SSDC).

Barrier   Anything used to control, prevent, or impede energy flows. Types of barriers include physical, equipment design, warning devices, procedures

System Safety for the 21ˢᵗ Century: The Updated and Revised Edition of System Safety 2000, by Richard A. Stephans
ISBN 0-471-44454-5   Copyright © 2004 John Wiley & Sons, Inc.

and work processes, knowledge and skills, and supervision. Barriers may be control or safety barriers or act as both (SSDC).

Component   As defined in NHB 5300.41(1D-2), "A combination of parts, devices, and structures, usually self-contained, which performs a distinctive function in the operation of the overall equipment. A 'black box' (e.g., transmitter, encoder, cryogenic pump, star tracker)" (NSTS 22254).

Contractor   A private sector enterprise or the organizational element of DoD or any other Government agency engaged to provide services or products within agreed limits specified by the MA (MIL-STD-882).

Corrective action   As defined in NHB 5300.41(1D-2), "Action taken to preclude occurrence of an identified hazard or to prevent recurrence of a problem" (NSTS 22254).

Critical item   A single failure point and/or a redundant element in a life- or mission-essential application where:

a. Redundant elements cannot be checked out during the normal ground turnaround sequence
b. Loss of a redundant element is not readily detectable in flight
c. All redundant elements can be lost by a single credible cause or event such as contamination or explosion.

Critical items list (CIL)   A listing comprised of all critical items identified as a result of performing the FMEA (NSTS 22254).

Criticality   The categorization of a hardware item by the worst case potential direct effect of failure of that item. In assigning hardware criticality, the availability of redundancy modes of operation is considered. Assignment of functional criticality, however, assumes the loss of all redundant hardware elements (NSTS 22254).

Category	Criticality Categories Definition
1	Loss of life or vehicle
1R	Redundant hardware element failure which could cause loss of life or vehicle
1S	Potential loss of life or vehicle due to failure of a safety or hazard monitoring system to detect, combat, or operate when required
2	Loss of mission; for Ground Support Equipment (GSE), loss of vehicle system
2R	Redundant hardware elements, the failure of which could cause loss of mission
3	All others

Damage   The partial or total loss of hardware caused by component failure; exposure of hardware to heat, fire, or other environments; human errors; or other inadvertent events or conditions (MIL-STD-882).

Documented safety analysis Documented safety analysis means a documented analysis of the extent to which a nuclear facility can be operated safely with respect to workers, the public, and the environment, including a description of the conditions, safe boundaries, and hazard controls that provide the basis for ensuring safety.

Failure As defined in NHB 5300.41(1D-2), "The inability of a system, subsystem, component, or part to perform its required function within specified limits, under specified conditions for a specified duration" (NSTS 22254).

Failure modes and effects analysis (FMEA) A systematic, methodical analysis performed to identify and document all identifiable failure modes at a prescribed level and to specify the resultant effect of the failure mode at various levels of assembly (NSTS 22254); the failure or malfunction of each system component is identified, along with the mode of failure (e.g., switch jammed in the "on" position). The effects of the failure are traced through the system and the ultimate effect on task performance is evaluated. Also called *failure mode* and *effect criticality analysis* (ASSE); a basic system safety technique wherein the kinds of failures that might occur and their effect on the overall product or system are considered. Example: The effect on a system by the failure of a single component, such as a register or a hydraulic valve (SSDC).

Fault tree An analytical tree used to determine fault. These may be used in accident/incident investigation or to determine accident potential before one has occurred (SSDC).

Graded approach Graded approach means the process of ensuring that the level of analysis, documentation, and actions used to comply with a requirement in this part are commensurate with

(1) The relative importance to safety, safeguards, and security;
(2) The magnitude of any hazard involved;
(3) The life cycle stage of a facility;
(4) The programmatic mission of a facility;
(5) The particular characteristics of a facility;
(6) The relative importance of radiological and nonradiological hazards; and
(7) Any other relevant factor.

Hazard A condition that is prerequisite to a mishap (MIL-STD-882); a condition or changing set of circumstances that presents a potential for injury, illness, or property damage. The potential or inherent characteristics of an activity, condition, or circumstance which can produce adverse or harmful consequences (ASSE); a condition that is prerequisite to a mishap (DODI 5000.36) (AFR 800-16); the presence of a potential risk situation caused by an unsafe act or condition (NSTS 22254); the potential for an energy flow(s) to result in an accident or otherwise adverse consequence (SSDC).

Hazard   Hazard means a source of danger (i.e., material, energy source, or operation) with the potential to cause illness, injury, or death to a person or damage to a facility or to the environment (without regard to the likelihood or credibility of accident scenarios or consequence mitigation).

Hazard analysis   The functions, steps, and criteria for design and plan of work, which identify hazards, provide measures to reduce the probability and severity potentials, identify residual risks, and provide alternative methods of further control (SSDC); a process of examining a system, design, or operation to discover inherent hazards, characterizing them as to level of risk and identifying risk-reduction alternatives (AFR 800-16); the determination of potential sources of danger and recommended resolutions in a timely manner for those conditions found in either the hardware/software systems, the person-machine relationship, or both, which cause loss of personnel capability, loss of system, or loss of life or injury to the public (NSTS 22254).

Hazard analysis techniques   Methods used to identify and evaluate hazards. These techniques cover the complete spectrum from qualitative preliminary hazard studies to system logic diagrams containing quantitative probabilities of mishap (AFR 800-16).

Hazard levels   The hazard level assigned to the identified hazard prior to applying the hazard reduction precedence sequence (HRPS) corrective action. Include hazard carried over for tracing from previous phases:

a. Catastrophic—No time or means are available for corrective action
b. Critical—May be counteracted by emergency action performed in a timely manner
c. Controlled—Has been countered by appropriate design, safety devices, alarm/caution and warning devices, or special automatic/manual procedures.

Hazardous event   An occurrence that creates a hazard (MIL-STD-882B).

Hazardous event probability   The likelihood, expressed in quantitative or qualitative terms, that a hazardous event will occur (MIL-STD-882B).

Hazard probability   The aggregate probability of occurrence of the individual hazardous events that create a specific hazard (MIL-STD-882).

Hazard report closure classification

a. Eliminated Hazard—A hazard that has been eliminated by removing the hazard source or by deleting the hazardous operations
b. Controlled Hazard—The likelihood of occurrence has been reduced to an acceptable level by implementing the appropriate hazard reduction precedence sequence to comply with program requirements

c. Accepted Risk—Hazard which has not been counteracted by redundancy, purge provisions, appropriate safety factors, containment/isolation provision, backup system/operation, safety devices, alarm/caution and warning devices, or special automatic/manual procedures. Catastrophic hazards, critical hazards, hazards resulting from failure to meet program requirements, and Single Failure Points (SFPs) in emergency systems will be documented. A hazard will be classified as an "accepted risk" only after (1) all reasonable risk avoidance measures have been identified, studied, and documented; (2) project/program management has made a decision to accept the risk on the basis of documented risk acceptance rationale; and (3) safety management has concurred in the accepted risk rationale (NSTS 22254).

Hazard report status

a. Closed—Corrective action to eliminate or control the hazard is completed, evaluated, and verified and management actions to accept the safety risks are completed. Actions taken, organization which performed actions and completion dates are to be documented in this data element.

b. Open—Corrective action evaluation and verification is in progress. The status shall remain open until management has reviewed the actions taken and accepted the safety risk. Actions required, organization documented in this data element (NSTS 22254).

Hazard severity   An assessment of the worst credible mishap that could be caused by a specific hazard (MIL-STD-882).

Human factors; human factors engineering   The application of the human biological and psychological sciences in conjunction with the engineering sciences to achieve the optimum mutual adjustment of man and his work, the benefits being measured in terms of human efficiency and well being. The principle disciplines involved are anthropometry, physiology, and engineering (SSDC).

Implementing command   The command or agency designated by HQ USAF to manage an acquisition program (AFR 800-2); AFR 800-4 includes modification programs (AFR 800-16).

Loss of personnel capability   As defined in NHB 5300.41(1D-2), "Loss of personnel function resulting in inability to perform normal or emergency operations. Also includes loss or injury to the public" (NSTS 22254).

Loss of vehicle system   As defined in NHB 5300.41(1D-2), "Loss of the capability to provide the level of system performance required for normal or emergency operations" (NSTS 22254).

Management oversight & risk tree (MORT)   A formal, disciplined logic or decision tree to relate and integrate a wide variety of safety concepts systematically. As an accident analysis technique, it focuses on three main concerns: specific oversights and omissions, assumed risks, and general management system weaknesses (SSDC).

Managing activity   The organizational element of DoD assigned acquisition management responsibility for the system, or prime or associate contractors or subcontractors who wish to impose system safety tasks on their suppliers (MIL-STD-882B).

Mishap   An unplanned event or series of events that results in death, injury, occupational illness, or damage to or loss of equipment or property (MIL-STD-882B); an unplanned event or series of events that result in death, injury, occupational illness, or damage to or loss of equipment or property, or damage to the environment. (DODI 5000.36) (AFR 800-16); a synonym for accident. Used by some government organizations, including NASA and DOD (SSDC).

Mission events   Time-oriented flight operations defined in flight checklists (NSTS 22254).

Near miss   An incident or an accident resulting in minor consequences although the potential for serious consequences was high (SSDC).

Nuclear facility   Nuclear facility means a reactor or a nonreactor nuclear facility where an activity is conducted for or on behalf of DOE and includes any related area, structure, facility, or activity to the extent necessary to ensure proper implementation of the requirements established by this Part.

Off-the-shelf item   An item determined by a material acquisition decision process review (DoD, Military Component, or subordinate organization as appropriate) to be available for acquisition to satisfy an approved material requirement with no expenditure of funds for development, modification, or improvement (e.g., commercial products, material developed by other Government agencies, or materiel developed by other countries). This item may be procured by the contractor or furnished to the contractor as Government-furnished equipment (GFE) or Government-furnished property (GFP) (MIL-STD-882B).

Operating and support hazard analysis (O&SHA)   As described in NHB 1700.1(V1-A) and this document. The O&SHA is to identify hazards and recommend risk reduction alternatives in procedurally controlled activities during all phases of intended use (NSTS 22254).

Preliminary hazard analysis (PHA)   As described in NHB 1700.1(V1-A) and this document. The PHA is to identify safety-critical areas, to identify and evaluate hazards, and to identify the safety design and operation requirements needed in the program concept phase (NSTS 22254).

Program manager   The single Air Force manager (system program director, program or project manager, or system, system program, or item manager) during any specific phase of acquisition life cycle (AFR 800-2) (AFR 800-16).

Quantified safety requirement A desired, predictable, and demonstrable level of safety, usually expressed as a mishap rate or probability of mishap (AFR 800-16).

Residual risk Risk remaining after the application of resources for prevention or mitigation (SSDC).

Risk Mathematically, expected loss; the probability of an accident multiplied by the quantified consequence of the accident (SSDC); an expression of the possibility of a mishap in terms of hazard severity and hazard probability (MIL-STD-882); note: Hazard exposure is sometimes included (AFR 800-16); as defined in NHB 5300.4(1D-2), "The chance (qualitative) of loss of personnel capability, loss of system, or damage to or loss of equipment or property" (NSTS 22254); a measure of both the probability and the consequence of all hazards of an activity or condition. A subjective evaluation of relative failure potential. In insurance, a person or thing insured (ASSE).

Risk acceptance The acceptance by an individual or organization of a level or degree of risk which has been identified as the potential consequence of a given course of action (ASSE).

Risk analysis The quantification of the degree of risk (SSDC).

Risk assessment The combined functions of risk analysis and evaluation (SSDC).

Risk assessment The amount or degree of potential danger perceived by a given individual when determining a course of action to accomplish a given task (ASSE).

Risk evaluation The appraisal of the significance or consequences of a given quantitative measure of risk (SSDC).

Risk management The process, derived through system safety principles, whereby management decisions are made concerning control and minimization of hazards of residual risks (SSDC); the professional assessment of all loss potentials in an organization's structure and operations, leading to the establishment and the administration of a comprehensive loss control program. Related to and dependent upon an ongoing program of accident prevention, risk management encompasses the selection of purchased insurance, self-insurance, and assumed risk. Its goal is to reduce losses to an acceptable minimum at the lowest possible cost (ASSE).

Risk visibility The documentation of a risk related to hardware, operations, procedures, software, and environment that provides Safety, project offices, and program management with the ability to evaluate accepted risks associated with planned operations (NSTS 22254).

Safe A condition wherein risks are as low as practicable and present no significant residual risk (SSDC).

Safety The control of accidental loss and injury (SSDC); freedom from those conditions that can cause death, injury, occupational illness, or damage to or loss of equipment or property (MIL-STD-882); a general term denoting an acceptable level of risk of, relative freedom from, and low probability of harm (ASSE); as defined in NHB 5300.4(1D-2), "Freedom from chance of injury or loss of personnel, equipment or property" (NSTS 22254).

Safety analysis   A systematic and orderly process for the acquisition and evaluation of specific information pertaining to the safety of a system (NSTS 22254).

Safety analysis (hazard analysis)   The entire complex of safety (hazard) analysis methods and techniques ranging from relatively informal job and task safety analyses to large complex safety analysis studies and reports (SSDC).

Safety analysis report (SAR)   A document prepared to document the results of a hazard analysis performed on a system, subsystem or operation. The specific minimum data elements for an SAR will be defined by data deliverable requirements for the program or project (NSTS 22254).

Safety basis   Safety basis means the documented safety analysis and hazard controls that provide reasonable assurance that a DOE nuclear facility can be operated safely in a manner that adequately protects workers, the public, and the environment.

Safety critical   As defined in NHB 5300.4(1D-2), "Facility, support, test, and flight systems containing:

a. Pressurized vessels, lines, and components
b. Propellants, including cryogenics
c. Hydraulics and pneumatics
d. High voltages
e. Radiation sources
f. Ordnance and explosive devices or devices used for ordnance and explosive checkout
g. Flammable, toxic, cryogenic, or reactive elements or compounds
h. High temperatures
i. Electrical equipment that operates in the area where flammable fluids or solids are located
j. Equipment used for handling program hardware
k. Equipment used for personnel walking and work platforms" (NSTS 22254).

Safety-critical computer software components   Those computer software components (processes, functions, values or computer program state) whose errors (inadvertent or unauthorized occurrence, failure to occur when required, occurrence out of sequence, occurrence in combination with other functions, or erroneous value) can result in a potential hazard, or loss of predictability or control of a system (MIL-STD-882).

Single failure point   As defined in NHB 5300.4(1D-2), "A single item of hardware, the failure of which would lead directly to loss of life, vehicle or mission. Where safety considerations dictate that abort be initiated when a redundant item fails, that element is also considered a single failure point" (NSTS 22254).

Space transportation system   An integrated sytem consisting of the Space Shuttle (Orbiter, External Tank [ET], Solid Rocket Booster [SRB], and flight kits), upper stages, Spacelab, and any associated flight hardware and software (NSTS 22254).

Subsystem   An element of a system that, in itself, may constitute a system (MIL-STD-882).

Subsystem hazard analysis (SSHA)   As described in NHB 1700.1(V1-A) and this document. The SSHA is to identify hazards to personnel, vehicle and other systems caused by loss of function, energy source, hardware failures, personnel action or inaction, software deficiencies, interactions of components within the subsystem, inherent design characteristics such as sharp edges, and incompatible materials, and environmental conditions such as radiation and sand (NSTS 22254).

Supporting command   The command assigned responsibility for providing logistics support; it assumes program management responsibility from the implementing command (AFR 800-2) (AFR 800-16).

System   A composite, at any level of complexity, of personnel, procedures, materials, tools, equipment, facilities, and software. The elements of this composite entity are used together in the intended operational or support environment to perform a given task or achieve a specific production, support, or mission requirement (MIL-STD-882); a set of arrangement of components so related or connected as to form a unity or organic whole. A set of facts, principles, rules, etc., classified or arranged in a regular, orderly form so as to show a logical plan linking the various parts. A method, plan, or classification. An orderly arrangement of interdependent activities and related procedures which implements and facilitates the performance of a major activity or organization. A set of components, humans or machines or both, which has certain functions and acts and interacts, one in relation to another, to perform some task or tasks in a particular environment or environments. Any configuration of elements in which the behavior properties of the whole are functions of both the nature of the elements and the manner in which they are combined (ASSE).

System analysis (safety analysis)   The formal analysis of a system and the interrelationships among its various parts (including plant and hardware, policies and procedures, and personnel) to determine the real and potential hazards within the system, and suggest ways to reduce and control those hazards.

System hazard analysis (SHA)   As described in NHB 1700.1(V1-A) and this document. The SHA is identical to the SSHA but at the system level. Once the subsystem levels have been established, a combination of subsystems comprise a system. In turn, a group of systems may comprise another system until the top system is identified (NSTS 22254).

System safety   The application of engineering and management principles, criteria, and techniques to optimize safety within the constraints of operational effectiveness, time, and cost throughout all phases of the system life cycle (DODI 5000.36) (AFR 800-16); an approach to accident prevention which involves the detection of deficiencies in system components which have an accident potential (ASSE); as defined in NHB 5300.4(1D-2), "The optimum degree of risk management within the constraints of operational effectiveness, time and cost attained through the application of management and engineering principles throughout all phases of a program (NSTS 22254); safety analysis (usually specialized and sophisticated) applied as an adjunct to design of an engineered system. While many associate system safety primarily with the hardware portion of the system, it includes all aspects of configuration control.

System safety analysis   The safety analysis of a complex process or system by means of a diagram or model that provides a comprehensive view of the process, including its principal elements and the ways in which they interrelate (ASSE).

System safety engineer   An engineer who is qualified by training and/or experience to perform system safety engineering tasks (MIL-STD-882).

System safety engineering   An engineering discipline requiring specialized professional knowledge and skills in applying scientific and engineering principles, criteria, and techniques to identify and eliminate hazards, or reduce the risk associated with hazards (MIL-STD-882).

System safety group/working group   A formally chartered group of persons, representing organizations associated with the system acquisition program, organized to assist the MA system program manager in achieving the system safety objectives. Regulations of the Military Components define requirements, responsibilities, and memberships (MIL-STD-882).

System safety management   An element of management that defines the system safety program requirements and ensures the planning, implementation and accomplishment of system safety tasks and activities consistent with the overall program requirements (MIL-STD-882).

System safety manager   A person responsible to program management for setting up and managing the system safety program (MIL-STD-882).

System safety program   The combined tasks and activities of system safety management and system safety engineering that enhance operational effectiveness by satisfying the system safety requirements in a timely, cost-effective manner throughout all phases of the system life cycle (MIL-STD-882).

System safety program plan   A description of the planned methods to be used by the contractor to implement the tailored requirements of this standard, including organizational responsibilities, resources, methods of accomplishment, milestones, depth of effort, and integration with other program engineering and management activities and related systems (MIL-STD-882).

User   Identified and authorized NASA, element contractor, or integration contractor personnel; flight crew equipment analyst; Orbiter experiments analyst; payload accommodations analyst; detailed secondary objective analyst; or RMS analyst (not inclusive) that have necessary access to the intercenter hazard data base system (NSTS 22254).

Additional sources of information for *System Safety 2000* were instructor's notes and student materials prepared by the author. These materials were developed for the following courses:

Systematic Occupational Safety (Practical MORT): 5-day versions taught to DOE, NASA, DOD, and private industry; 1-day version conducted as professional development seminar at National Safety Congresses

Systematic Accident Investigation: 8-, 5-, 3-, and 1-day versions taught to DOE, NASA, and private industry

Accident Analysis: 1-day professional development seminar presented at National Safety Congress

You and System Safety: 1-day professional development seminar presented at National Safety Congresses and to private industry

System Safety Workshop: 5- and 3-day versions presented to NASA and private industry

Facility System Safety Workshop: 5-day course developed for and taught to DOD

Overview of Facility System Safety for the Architect/Engineer: 3-day course developed for DOD and private industry

Combat Oriented Mishap Prevention Analysis System (COMPAS): 8-day course developed for and taught to DOD

Techniques of Safety Engineering Analysis: semester course developed and taught at the University of Houston—Clear Lake

Advanced Occupational Safety Course (AOSC): 15-day course developed and taught for DOD

Briscoe, G. J. 1977. *Risk Management Guide* (ERDA 76-45/11; SSDC-11). Idaho Falls, ID: Energy Research and Development Administration.

Browning, R. L. 1980. *The Loss Rate Concept in Safety Engineering.* New York: Marcel Dekker.

Buys, J. R. 1977. *Standardization Guide for Construction and Use of MORT-Type Analytical Trees* (ERDA 76-45/8; SSDC-8). Idaho Falls, ID: Energy Research and Development Administration.

Clements, P. L. 1987. *Concepts in Risk Management.* Sverdrup Technology, Inc.

System Safety for the 21ˢᵗ Century: The Updated and Revised Edition of System Safety 2000,
by Richard A. Stephans
ISBN 0-471-44454-5   Copyright © 2004 John Wiley & Sons, Inc.

Crosetti, P. A. 1982. *Reliability and Fault Tree Analysis Guide* (DOE 76-45/22; SSDC-22). Idaho Falls, ID: Department of Energy.

Dhillon, B. S., and Singh, C. 1981. *Engineering Reliability: New Techniques and Applications.* New York: John Wiley & Sons.

Fussell, J. B., and Burdick, G. R. 1977. *Nuclear Systems Reliability Engineering and Risk Assessment.* Society for Industrial and Applied Mathematics.

Goldwaite, William H., and others. 1985. *Guidelines for Hazard Evaluation Procedures.* New York: American Institute of Chemical Engineers.

Hammer, Willie. 1972. *Handbook of System and Product Safety.* Englewood Cliffs, NJ: Prentice-Hall.

Henley, E. J., and Kumamoto, H. 1981. *Reliability Engineering and Risk Assessment.* Englewood Cliffs, NJ: Prentice-Hall.

Henley, E. J., and Kumamoto, H. 1985. *Designing for Reliability and Safety Control.* Englewood Cliffs, NJ: Prentice-Hall.

Johnson, William G. 1973. *MORT, the Management Oversight and Risk Tree.* Washington, DC: U.S. Atomic Energy Commission.

Johnson, William G. 1980. *MORT Safety Assurance Systems.* New York: Marcel Dekker.

McCormick, E. J. 1976. *Human Factors in Engineering and Design.* New York: McGraw Hill.

McElroy, F. E., ed. 1981. *Accident Prevention Manual for Industrial Operations.* 2 vols. Chicago: National Safety Council.

Malasky, S. W. 1982. *System Safety: Technology and Application.* New York: Garland Press.

National Aeronautics and Space Administration. 1970. *System Safety.* NHB 1700.1(V3). Washington, DC: Safety Office, NASA.

National Aeronautics and Space Administration. 1987. *Methodology for Conduct of NSTS Hazard Analyses.* NSTA 22254. Houston: NSTS Program Office, Johnson Space Center.

National Aeronautics and Space Administration. n.d. *Safety, Reliability, and Quality Assurance Phase I & II Training Manual.* Houston, Office of SR&QA, Johnson Space Center.

Rogers, W. P. 1971. *Introduction to System Safety Engineering.* New York: John Wiley & Sons.

Roland, H. E., and Moriarty, Brian. 1981. *System Safety Engineering and Management.* New York: John Wiley & Sons.

U.S. Air Force. 1987. *System Safety Handbook for the Acquisition Manager.* SDP 127-1. Los Angeles: HA Space Division/SE.

U.S. Army. Facility System Safety Manual Washington, DC: HQ Safety, USACE. Draft.

U.S. Department of Defense. 1984 (updated by Notice 1, 1987). *System Safety Program Requirements.* MIL-STD-882B. Washington, DC: Department of Defense.

Vesely, W. E., and others. 1981. *Fault Tree Handbook.* NUREG-0492. Washington, DC: U.S. Government Printing Office.

The following is a listing of pamphlets prepared by the System Safety Development Center staff, Idaho Falls, Idaho:

SSDC 1   *Occupancy Use Manual*
SSDC 2   *Human Factors in Design*
SSDC 3   *A Contractor Guide to Advance Preparation for Accident Investigation*
SSDC 4   *MORT User's Manual*
SSDC 5   *Reported Significant Observation (RSO) Studies*
SSDC 6   *Training as Related to Behavioral Change*
SSDC 7B   *DOE Guide to the Classification of Recordable Accidents*
SSDC 8   *Standardization Guide for Construction and Use of MORT Type Analytic Trees*
SSDC 9   *Safety Information System Guide*
SSDC 10   *Safety Information System Cataloging*
SSDC 11   *Risk Management Guide*
SSDC 12   *Safety Considerations in Evaluation of Maintenance Programs*
SSDC 13   *Management Factors in Accident/Incidents* (Including Management Self-Evaluation Checksheets)
SSDC 14   *Events and Causal Factors Charting*
SSDC 15   *Work Process Control Guide*
SSDC 16   *SPRO Drilling and Completion Operations*
SSDC 17   *Applications of MORT to Review of Safety Analyses*
SSDC 18   *Safety Performance Measurement System*
SSDC 19   *Job Safety Analysis*
SSDC 20   *Management Evaluation and Control of Release of Hazardous Materials*
SSDC 21   *Change Control and Analysis*
SSDC 22   *Reliability and Fault Tree Analysis Guide*
SSDC 23   *Safety Appraisal Guide*
SSDC 24   *Safety Assurance System Summary (SASS) Manual for Appraisal*
SSDC 25   *Effective Safety Review*
SSDC 26   *Construction Safety Monographs (not available for general distribution)*
SSDC 27   *Accident/Incident Investigation Manual (2d edition)*
SSDC 28   *Glossary of SSDC Terms and Acronyms*
SSDC 29   *Barrier Analysis*
SSDC 30   *Human Factors Management*
SSDC 31   *The Process of Task Analysis*
SSDC 32   *The Impact of the Human on System Safety Analysis*
SSDC 33   *The MORT Program and the Safety Performance Measurement System*
SSDC 34   *Basic Human Factors Considerations*
SSDC 35   *A Guide for the Evaluation of Displays*
SSDC 36   *MORT-Based Safety Professional/Program Development and Improvement Time/Loss Analysis*

SSDC 37   *Time/Loss Analysis*
SSDC 38   *Safety Considerations for Security Programs*
SSDC 39   *Process Operational Readiness and Operational Readiness Followon*
SSDC 40   *The Assessment of Behavioral Climate*
SSDC 41   *Investigating and Reporting Accidents Effectively*
MORT Charts

Information about SSDC pamphlets may be obtained from Systems Safety Development Center, EG&G Idaho, Inc., P.O. Box 1625, Idaho Falls, ID 83415-3405.

For information about courses and materials prepared by the author, including mini-MORT and PET charts, contact Joe Stephenson, 8429 Honeywood Circle, Las Vegas, NV 89128.

System Safety for the 21ˢᵗ Century: The Updated and Revised Edition of System Safety 2000,
by Richard A. Stephans
ISBN 0-471-44454-5   Copyright © 2004 John Wiley & Sons, Inc.